A Level
Biology
for OCR

A

Michael Fisher

OXFORD
UNIVERSITY PRESS

Great Clarendon Street, Oxford, OX2 6DP, United Kingdom

Oxford University Press is a department of the University of Oxford. It furthers the University's objective of excellence in research, scholarship, and education by publishing worldwide. Oxford is a registered trade mark of Oxford University Press in the UK and in certain other countries

British Library Cataloguing in Publication Data
Data available

978-0-19-835194-8

10 9

Paper used in the production of this book is a natural, recyclable product made from wood grown in sustainable forests. The manufacturing process conforms to the environmental regulations of the country of origin.

Printed in Great Britain by CPI Group (UK) Ltd., Croydon CR0 4YY

Acknowledgements

Cover: Valentina Razumova / Shutterstock
Artwork by Q2A Media

This book has been written to support students studying for OCR A Level Biology A. It covers the A Level modules from the specification. The modules covered are shown in the contents list, which also shows you the page numbers for the main topics within each module.

AS exam

A level exam

Year 1 content

1 Development of practical skills in biology
2 Foundations in biology
3 Exchange and transport
4 Biodiversity, evolution, and disease

Year 2 content

5 Communication, homeostasis, and energy
6 Genetics, evolution, and ecosystems

A Level exams will cover content from Year 1 and Year 2 and will be at a higher demand.

Module 2 Foundations in Biology

Chapter 2 Basic components of living systems   2
2.1	Microscopy	2
2.2	Magnification and calibration	3
2.3	More microscopy	4
2.4	Eukaryotic cell structure	5
2.5	The ultrastructure of plant cells	7
2.6	Prokaryotic and eukaryotic cells	8
	Practice questions	9

Chapter 3 Biological molecules   10
3.1	Biological elements	10
3.2	Water	11
3.3	Carbohydrates	12
3.4	Testing for carbohydrates	14
3.5	Lipids	15
3.6	Structure of proteins	16
3.7	Types of proteins	18
3.8	Nucleic acids	19
3.9	DNA replication and the genetic code	21
3.10	Protein synthesis	22
3.11	ATP	23
	Practice questions	24

Chapter 4 Enzymes   25
4.1	Enzyme action	25
4.2	Factors affecting enzyme activity	26
4.3	Enzyme inhibitors	28
4.4	Cofactors, coenzymes, and prosthetic groups	29
	Practice questions	30

Chapter 5 Plasma membranes   31
5.1	The structure and function of membranes	31
5.2	Factors affecting membrane structure	33
5.3	Diffusion	34
5.4	Active transport	35
5.5	Osmosis	37
	Practice questions	38

Chapter 6 Cell division   39
6.1	The cell cycle	39
6.2	Mitosis	41
6.3	Meiosis	43
6.4	The organisation and specialisation of cells	45
6.5	Stem cells	47
	Practice questions	48

Module 3 Exchange and transport

Chapter 7 Exchange surfaces and breathing   49
7.1	Specialised exchange surfaces	49
7.2	The mammalian gaseous exchange system	50
7.3	Measuring the process	50
7.4	Ventilation and gas exchange in other organisms	53
	Practice questions	55

Chapter 8 Transport in animals   56
8.1	Transport systems in multicellular animals	56
8.2	Blood vessels	58
8.3	Blood, tissue fluid, and lymph	60
8.4	Transport of oxygen and carbon dioxide in the blood	62
8.5	The heart	64
	Practice questions	66

Chapter 9 Transport in plants   67
9.1	Transport systems in dicotyledonous plants	67
9.2	Water transport in multicellular plants	69
9.3	Transpiration	70
9.4	Translocation	72
9.5	Plant adaptations to water availability	73
	Practice questions	74

Module 4 Biodiversity, evolution, and disease

Chapter 10 Classification and evolution   75
10.1	Classification	75
10.2	The five kingdoms	76
10.3	Phylogeny	77
10.4	Evidence for evolution	78
10.5	Types of variation	79
10.6	Representing variation graphically	81
10.7	Adaptations	84
10.8	Changing population characteristics	86
	Practice questions	88

Chapter 11 Biodiversity   89
11.1	Biodiversity	89
11.2	Types of sampling	90
11.3	Sampling techniques	91
11.4	Calculating biodiversity	92
11.5	Calculating genetic biodiversity	93
11.6	Factors affecting biodiversity	95
11.7	Reasons for maintaining biodiversity	96
11.8	Methods of maintaining biodiversity	97
	Practice questions	98

Chapter 12 Communicable diseases   99
12.1	Animal and plant pathogens	99
12.2	Animal and plant diseases	100
12.3	The transmission of communicable diseases	101
12.4	Plant defences against pathogens	102
12.5	Non-specific animal defences against pathogens	103
12.6	The specific immune system	104
12.7	Preventing and treating disease	106
	Practice questions	107

Module 5 Communication, homeostasis, and energy

Chapter 13 Neuronal communication   108
13.1	Coordination	108
13.2	Neurones	109

13.3	Sensory receptors	110
13.4	Nervous transmission	111
13.5	Synapses	113
13.6	Organisation of the nervous system	114
13.7	Structure and function of the brain	114
13.8	Reflexes	115
13.9	Voluntary and involuntary muscles	116
13.10	Sliding filament model	116
	Practice questions	118

Chapter 14 Hormonal communication   119
14.1	Hormonal communication	119
14.2	The pancreas	120
14.3	Regulation of blood glucose concentration	121
14.4	Diabetes and its control	122
14.5	Coordinated responses	124
14.6	Controlling heart rate	125
	Practice questions	127

Chapter 15 Homeostasis   128
15.1	The principles of homeostasis	128
15.2	Thermoregulation in ectotherms	130
15.3	Thermoregulation in endotherms	130
15.4	Excretion, homeostasis, and the liver	132
15.5	The structure and function of the mammalian kidney	133
	Practice questions	137

Chapter 16 Plant responses   138
16.1	Plant hormones and growth in plants	138
16.2	Plant responses to abiotic stress	139
16.3	Plant responses to herbivory	139
16.4	Tropisms in plants	140
16.5	The commercial use of plant hormones	141
	Practice questions	142

Chapter 17 Energy for biological processes   143
17.1	Energy cycles	143
17.2	ATP synthesis	144
17.3	Photosynthesis	145
17.4	Factors affecting photosynthesis	149
	Practice questions	150

Chapter 18 Respiration   151
18.1	Glycolysis	151
18.2	Linking glycolysis and the Krebs cycle	152
18.3	The Krebs cycle	153
18.4	Oxidative phosphorylation	154
18.5	Anaerobic respiration	155
18.6	Respiratory substrates	156
	Practice questions	157

Module 6 Genetics, evolution, and ecosystems

Chapter 19 Genetics of living systems   158
19.1	Mutations and variation	158
19.2	Control of gene expression	160

19.3	Body plans	162
	Practice questions	163

Chapter 20 Patterns of inheritance and variation   164
20.1	Variation and inheritance	164
20.2	Monogenic inheritance	165
20.3	Dihybrid inheritance	167
20.4	Phenotypic ratios	168
20.5	Evolution	170
20.6	Speciation and artificial selection	172
	Practice questions	174

Chapter 21 Manipulating genomes   175
21.1	DNA profiling	175
21.2	DNA sequencing and analysis	177
21.3	Using DNA sequencing	177
21.4	Genetic engineering	179
21.5	Gene technology and ethics	180
	Practice questions	181

Chapter 22 Cloning and biotechnology   182
22.1	Natural cloning in plants	182
22.2	Artificial cloning in plants	182
22.3	Cloning in animals	184
22.4	Microorganisms and biotechnology	186
22.5	Microorganisms, medicines, and bioremediation	186
22.6	Culturing microorganisms in the laboratory	187
22.7	Culturing microorganisms on an industrial scale	187
22.8	Using immobilised enzymes	189
	Practice questions	190

Chapter 23 Ecosystems   191
23.1	Ecosystems	191
23.2	Biomass transfer through an ecosystem	191
23.3	Recycling within ecosystems	194
23.4	Succession	196
23.5	Measuring the distribution and abundance of organisms	197
	Practice questions	198

Chapter 24 Populations and sustainability   199
24.1	Population size	199
24.2	Competition	200
24.3	Predator–prey relationships	200
24.4	Conservation and preservation	201
24.5	Sustainability	201
24.6 – 24.9	Ecosystem management	202
	Practice questions	203

Answers to practice questions	**204**
Answers to summary questions	**210**
Appendix	**234**

This book contains many different features. Each feature is designed to support and develop the skills you will need for your examinations, as well as foster and stimulate your interest in biology.

Worked example
Step-by-step worked solutions.

Common misconception
Common student misunderstandings clarified.

Go further
Familiar concepts in an unfamiliar context.

Maths skills
A focus on maths skills.

Question and model answer
Sample answers to exam style questions.

Practical skill
Support for the practical knowledge requirements of the exam.

Summary Questions

1 These are short questions at the end of each topic.

2 They test your understanding of the topic and allow you to apply the knowledge and skills you have acquired.

3 The questions are ramped in order of difficulty.

Specification references
→ At the beginning of each topic, there are specification references to allow you to monitor your progress.

Key term
Pulls out key terms for quick reference.

Revision tip
Prompts to help you with your understanding and revision.

Synoptic link
These highlight the key areas where topics relate to each other. As you go through your course, knowing how to link different areas of biology together becomes increasingly important. Many exam questions, particularly at A Level, will require you to bring together your knowledge from different areas.

Chapter 13 Practice questions

1   Which of the following statements is/are true of the functions of the sympathetic nervous system?   (*1 mark*)

   1 Increases heart rate.

   2 Increases the speed at which food moves through the gut.

   3 Decreases sweat production.

   A 1, 2, and 3 are correct

   B Only 1 and 2 are correct

   C Only 2 and 3 are correct

   D Only 1 is correct

2   What is the function of the cerebellum?   (*1 mark*)

   A Coordination of balance and muscular movement.

   B Controls conscious thought processes.

   C Controls body temperature.

   D Regulates heart rate.

3   Which of the following is/are present in the membrane of an acetylcholinergic presynaptic neurone?   (*1 mark*)

   1 Calcium ion channel

   2 Acetylcholine receptor

   3 Acetylcholinesterase

   A 1, 2, and 3 are correct

   B Only 1 and 2 are correct

   C Only 2 and 3 are correct

   D Only 1 is correct

4   Which of the following statements is/are true of the role of calcium ions in skeletal muscle contraction?   (*1 mark*)

   1 Calcium ions are released into the sarcoplasm.

   2 Calcium ions bind to tropomyosin.

   3 Calcium ions inactivate an enzyme called myosin kinase.

   A 1, 2, and 3 are correct

   B Only 1 and 2 are correct

   C Only 2 and 3 are correct

   D Only 1 is correct

5   Which of the following statements is/are true of the changes in a sarcomere during muscle contraction?   (*1 mark*)

   1 Myosin filaments shorten.

   2 The H zone shortens.

   3 I band shortens.

   A 1, 2, and 3 are correct

   B Only 1 and 2 are correct

   C Only 2 and 3 are correct

   D Only 1 is correct

6   Outline the differences between the sympathetic nervous system and the parasympathetic nervous system.   (*6 marks*)

2.1 Microscopy

Specification reference: 2.1.1 (a), (b), (c), and (d)

This chapter is concerned with the appearance of cells. Observing biological material is a fundamental scientific skill. The invention of the light microscope more than 400 years ago allowed cells to be observed. Since then, microscope technology has advanced to enable fine details to be seen. Here you will learn how a modern light microscope is used.

How does a light microscope work?

Light shines up though the sample being observed and through two lenses (**objective** and **eyepiece**). Each lens magnifies the image; the combined effect of the two lenses can increase an image by up to 2000 times its actual size.

> ### Revision tip: The condenser lens
> Light first passes through a condenser lens. This does not magnify; it focuses light on the sample being observed. You are not required to learn about condenser lenses.

⚛ Practical skill: Preparation of a sample

The type of biological material being studied will dictate how it is prepared for a light microscope. Specimens may need to be **sectioned** (cut into thin slices), **fixed** (preserved and sterilised), and **stained** (see below). Cover slips can be placed over specimens immersed in liquid (**wet mounting**) or without liquid (**dry mounting**).

Staining improves the visibility of structures within a specimen. Stains increase the contrast between different structures (because components take up stains in different ways). **Differential staining** can therefore distinguish between structures in cells or between different species (if their cells contain different components). For example, some bacteria retain crystal violet and appear purple (*Gram-positive bacteria*), whereas others do not (*Gram-negative bacteria*).

Revision tip: Drawing conclusions
Scientific drawings should be clear and precise; they should be produced using a 'less-is-more' principle (e.g. drawing smooth continuous lines, without unnecessary detail, and showing essential features).

Summary questions

1   Explain why a specimen needs to be sectioned before it is examined under a light microscope. *(1 mark)*

2   Describe the roles of the eyepiece and objective lenses in a light microscope. *(2 marks)*

3   FABIL stains lignin yellow and cellulose blue. Using knowledge from Topic 2.5, The ultrastructure of plant cells, and Topic 9.1, Transport systems in dicotyledonous plants, suggest the benefit of FABIL to a person observing plant tissue. *(3 marks)*

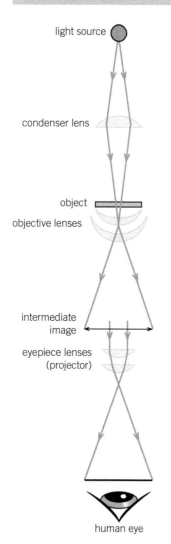

▲ **Figure 1** *The path of light through a microscope*

2.2 Magnification and calibration
Specification reference: 2.1.1(b), (e), and (f)

When observing biological material under a microscope, scientists need to know how much larger the image is in comparison to the actual size of the material. This is the concept of magnification, which we discuss here.

Magnification and resolution

Magnification is the extent to which the actual size of an object is enlarged into the image seen through a microscope. **Resolution** is the extent to which two objects can be distinguished as separate structures. More detail can be seen at higher resolutions.

Maths skill: Calculating magnification

$$\text{Magnification} = \frac{\text{size of image}}{\text{actual size of object}}$$

Worked example: Rearranging the formula

You may be asked to calculate image size or the actual size of an object, rather than magnification. This requires the formula to be rearranged. For example, if a microscope has a magnification of ×1500, and a red blood cell appears to have a diameter of 1.05 cm, how do we calculate the actual diameter of the cell?

$$\text{Actual size} = \frac{\text{size of image}}{\text{magnification}} = \frac{1.05}{1500} = 0.0007 \text{ cm } (7 \,\mu m)$$

Revision tip: If you forget the formula ...

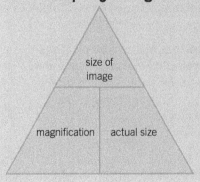

Mnemonics are useful tools for remembering the arrangement of formula triangles. For example, you can use 'I'm Mag actually' to remember the arrangement of the magnification formula.

Practical skill: Calibrating microscopes

A microscope eyepiece contains a scale without divisions (i.e. a graticule). Calibrating a microscope involves working out the length represented by each division in this scale at a particular magnification. This enables you to measure the sample being observed.

The graticule scale is compared to a stage micrometer (a slide with a scale in µm). You can then calculate the number of µm per eyepiece division. For example, each division on the Figure 1 eyepiece scale represents 30 µm (12 stage micrometer units = 120 µm, divided by four eyepiece units = 30 µm).

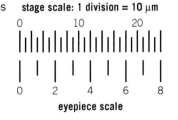

stage scale: 1 division = 10 µm

eyepiece scale

▲ **Figure 1** *Graticule and micrometer scales are compared*

Key term

Resolution: The shortest distance between two objects that are still seen as separate structures.

Revision tip: It can all be a blur

Magnification may increase without resolution improving. This will result in a larger image, but one that is blurred.

Revision tip: How significant?

Check the number of significant figures required in an answer. A calculator answer of 5.478 µm would be 5.5 µm to two significant figures, for example.

Summary questions

1 State the difference between magnification and resolution. (*2 marks*)

2 Calculate the diameter of a bacterium that appears 2.4 mm wide under a microscope (×1600 magnification). (*2 marks*)

3 An erythrocyte has a diameter of 7 µm. Calculate the image size when viewed at magnification ×1600 under a light microscope. Express your answer in standard form and to two significant figures. (*3 marks*)

2.3 More microscopy

Specification reference: 2.1.1(a) and (f)

The study of cell details was made possible by the invention of the light microscope. Nowadays, scientists have access to microscopes that are much more powerful. Rather than using regular visible light, these microscopes use electron beams and lasers, allowing a wealth of detail to be visualised within cells.

Comparing microscopes

Electron beams have shorter wavelengths than light waves. This is why electron microscopes have greater resolutions than light microscopes. Electron microscopes work by either transmitting electrons through a sample (TEM) or scanning the sample (SEM). The two microscopes have different properties, as shown in the following table.

Revision tip: Artefacts

An artefact is something visible in a microscope image that is not a natural part of the specimen (e.g. structural distortion created during preparation).

Synoptic link

See Topic 2.1, Microscopy, for a reminder of the approach to scientific drawing.

	Light	Electron microscope	
		Transmission (TEM)	Scanning (SEM)
How does it work?	See Topic 2.1, Microscopy	Electron beams pass through the specimen	Electrons are reflected back from the specimen and detected
Magnification	Up to ×2000	Up to ×500 000	Up to ×100 000
Resolution (nm)	200	0.5	3–10
Benefits	Inexpensive, with short preparation time. Colours and living material can be observed	High magnification and resolution. Enables intracellular details to be observed	High magnification and resolution. 3D images
Disadvantages	Limited magnification and resolution	Expensive, with complex preparation of samples. Images are black and white (but artificial colour can be added). Artefacts can be introduced during preparation.	

A modern light microscope

The confocal laser scanning microscope (CLSM) uses laser light to scan a specimen. Light is absorbed by fluorescent chemicals and radiated back from the specimen. A laser is a device that produces light in which the waves are lined up together in the same direction. As a result, light from lasers is very bright and can be focused on a small spot.

Although the resolution of these microscopes is poor compared to electron microscopes, they have two benefits:

- *3D images* are produced
- They are *non-invasive*, which allows living tissue to be observed.

Summary questions

1   Describe and explain the difference in the resolving power of electron and light microscopes.   *(2 marks)*

2   A chloroplast with a diameter of 6 µm was magnified using a transmission electron microscope. The magnification was set to ×50 000. Calculate the size of the image. Express your answer in standard form.   *(2 marks)*

3   Evaluate the advantages and disadvantages of electron microscopes and CLSMs.   *(5 marks)*

 Practical skill: Drawing from electron micrographs

Electron microscopes show more detail than light microscopes because of their greater resolution. The principles you should follow when drawing images are the same in either context.

2.4 Eukaryotic cell structure

Specification reference: 2.1.1(g), (i), and (j)

Thanks to microscopy, scientists now have an excellent understanding of the contents and structure of cells. The architecture of cells, however, can differ between species. You will examine some of these differences in the following three topics. We begin by discussing eukaryotic cells.

Cellular components of eukaryotic cells

Species consist of either prokaryotic or eukaryotic cells (see Topic 2.6, Prokaryotic and eukaryotic cells). Animals, fungi, plants, and protoctista are eukaryotes. The most significant feature of eukaryotic cells is that they contain compartments called **organelles**.

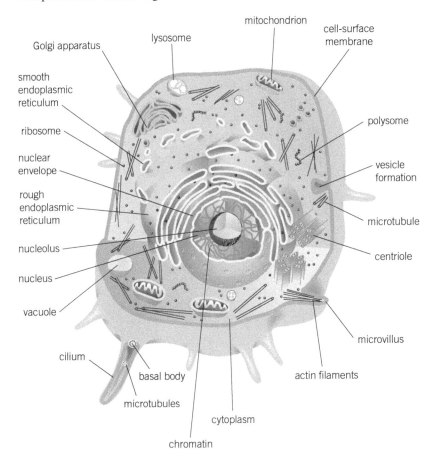

▲ **Figure 1** *The generalised ultrastructure of an animal eukaryotic cell*

The following table outlines the functions and features of membrane-bound organelles and cell components lacking membranes in eukaryotic cells.

Synoptic link

You will learn more about the classification and naming of organisms in Topic 10.2, The five kingdoms.

Revision tip: Ultrastructure

The ultrastructure of a cell is the fine detail that can be observed only by using an electron microscope.

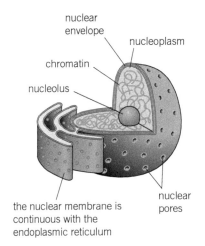

the nuclear membrane is continuous with the endoplasmic reticulum

▲ **Figure 2** *The relationship between the nucleus and endoplasmic reticulum*

Key term

Organelles: Structures in eukaryotic cells with specific roles. (They are often defined as being membrane-bound, although some scientists include structures without membranes.)

Component	Function	Key features
Nucleus (plural: nuclei)	Contains genetic information	Surrounded by a **nuclear envelope** (containing pores)
		Contains **chromatin** (DNA + histone proteins), which condenses to form chromosomes
		Contains a **nucleolus** (which produces rRNA)
Mitochondria (singular: mitochondrion)	ATP production through aerobic **respiration**	Double membrane; the inner membrane folds to form **cristae**
		Internal fluid is called the **matrix**

Basic components of living systems

Revision tip: Cytosol vs cytoplasm

Cytosol is the fluid within a cell. Cytoplasm comprises the cytosol plus the organelles suspended within it. The one exception is the nucleus, which is considered to be separate to the cytoplasm.

Synoptic link

Chapter 3, Biological molecules, delves further into DNA structure and protein synthesis.

Lysosomes	Breaking down **waste** (e.g. old organelles)	Specialised **vesicles** (membranous sacs), containing **hydrolytic enzymes**
Smooth endoplasmic reticulum (ER)	**Lipid** and carbohydrate synthesis	Flattened membrane-bound sacs (**cisternae**)
Rough endoplasmic reticulum (ER)	**Protein synthesis**	Cisternae bound to **ribosomes** (which are made of RNA; ribosomes can also appear loose in cytoplasm)
Golgi apparatus	**Modifying proteins** and packaging them into vesicles	Flattened sacs (similar to smooth ER)
Cytoskeleton	Maintaining cell shape Control of cell movement and of organelle movement in cells Compartmentalisation of organelles	**Microfilaments** (made from actin) control cell movement and cytokinesis (see Topic 6.2, Mitosis). **Microtubules** (made from tubulin) regulate shape and organelle movements. They form centrioles and spindle fibres (see Topic 6.2, Mitosis).
Flagella (singular: flagellum) and cilia (singular: cilium)	Flagella enable cell movement. Cilia move substances across cell surfaces.	Both comprise a cylinder containing 11 **microtubules** (nine in a circle and two in the centre). Flagella are longer than cilia.
Cell surface (plasma) membrane	See Topic 5.1, The structure and function of membranes	

Summary questions

1 State one structural similarity and one structural difference between a flagellum and a cilium. *(2 marks)*

2 Describe the stages undergone by a protein from its production to its release from a cell. *(5 marks)*

3 Actin microfilaments can form networks that assemble and disassemble rapidly within cells. Suggest how this activity benefits some cells. *(3 marks)*

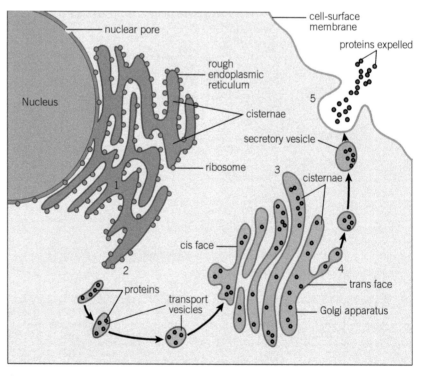

▲ **Figure 3** *Protein production and processing: (1) Synthesis in rough ER. (2) Transport in vesicles. (3) Structural modification in the Golgi apparatus. (4) Modified proteins leave the Golgi in vesicles. (5) Some proteins are released from the cell (others are retained).*

2.5 The ultrastructure of plant cells

Specification reference: 2.1.1(g)

Plant cell features

Plant cells are eukaryotic and tend to possess the components we discussed in Topic 2.4, Eukaryotic cell structure (although centrioles are absent from flowering plants). Plants, in addition, have three structures that other eukaryotic cells lack: cellulose cell walls, large vacuoles, and chloroplasts.

Component	Description	Function
Cellulose cell wall	A strong barrier on the outside of the cell membranes. It is composed of a polysaccharide called **cellulose**.	Provides shape and rigidity
Large, permanent vacuole	A large, central sac (surrounded by a membrane known as a **tonoplast**). It principally contains water.	Increases **turgor** (i.e. the extent to which cells are filled with water)
Chloroplasts	An organelle containing fluid (**stroma**) and a network of membranes (**thylakoids**, which are stacked as **grana**).	The organelle responsible for **photosynthesis**

Synoptic link

You will learn about the structure of cellulose in Topic 3.3, Carbohydrates.

Revision tips: Walls and vacuoles in other organisms

Cell walls are found in organisms other than plants. Fungi possess complex walls containing chitin. Bacteria have walls made of peptidoglycan.

All fungal cells have at least one vacuole, although they display some functional differences to plant vacuoles. Some bacterial cells contain vacuoles.

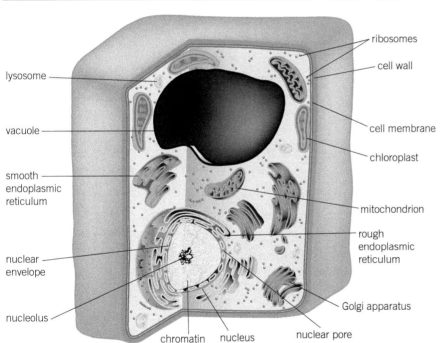

▲ **Figure 1** *A generalised plant cell*

Labels (Figure 1): ribosomes, cell wall, cell membrane, chloroplast, mitochondrion, rough endoplasmic reticulum, Golgi apparatus, nuclear pore, nucleus, chromatin, nucleolus, nuclear envelope, smooth endoplasmic reticulum, vacuole, lysosome

Labels (Figure 2): chloroplast envelope comprising inner and outer membranes, starch grain – stores photosynthetic products, ribosomes – small (70S) type, oil droplet, stroma, thylakoid – stacked together to form a granum. Containing photosynthetic pigments, large thylakoid, connecting grana (= intergranal lamella), chloroplast DNA – circular formation

▲ **Figure 2** *Chloroplast structure*

Summary questions

1 State the principal role of a vacuole in plant cells.   (*1 mark*)

2 *Chloroplasts are the only truly unique feature of plant cells.* Evaluate the validity of this statement.   (*2 marks*)

3 Ribosomes in chloroplasts are similar to the 70S ribosomes of prokaryotic cells, but with some structural differences. What does this suggest about the possible evolutionary history of chloroplasts?   (*3 marks*)

Revision tips: Standard advice

The ability to convert measurements between decimal and standard form is an important skill. The following examples provide a good comparison.

Chloroplast length
$8\,\mu m = 0.000008\,m = 8 \times 10^{-6}\,m$ (standard form)

Palisade cell diameter
$40\,\mu m = 0.000040\,m = 4 \times 10^{-5}\,m$ (standard form)

2.6 Prokaryotic and eukaryotic cells

Specification reference: 2.1.1(k)

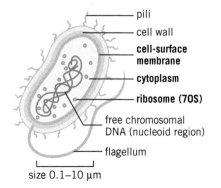

- pili
- cell wall
- **cell-surface membrane**
- **cytoplasm**
- **ribosome (70S)**
- free chromosomal DNA (nucleoid region)
- flagellum

size 0.1–10 μm

▲ **Figure 1** *A generalised prokaryotic cell; the features in bold are found in all eukaryotic cells as well*

Revision tip: S for ... size?

S is a unit that indicates how quickly a substance settles at the bottom of a centrifuge tube (centrifugation is a method for separating cell contents). In terms of ribosomes, we can think of *S* as a size indicator (i.e. 70S ribosomes are smaller than 80S).

Summary questions

1 State the difference in the structure of prokaryotic and eukaryotic cell walls. *(2 marks)*

2 Bacteria possess pili, which are hair-like structures that allow the exchange of genetic material between cells. Suggest why this is useful for bacteria. *(2 marks)*

3 Describe the differences in DNA packaging between prokaryotic and eukaryotic cells. *(4 marks)*

You learned about the ultrastructure of eukaryotic cells in Topic 2.4, Eukaryotic cell structure, and Topic 2.5, The ultrastructure of plant cells. Here you will learn about prokaryotic cells and examine the differences between the two categories of cell.

How are prokaryotic cells different?

Prokaryotes are always unicellular, whereas many eukaryotic species are composed of more than one cell.

These two types of cell also exhibit many structural differences, as shown in the following table.

Feature	Prokaryotic cells	Eukaryotic cells
Membrane-bound organelles	No	Yes
DNA	One main molecule of DNA (**circular** and found naked in the cytoplasm) Additional DNA can be found in **plasmids**	Linear (**chromosomes**) and associated with histone proteins in a nucleus Additional DNA found in mitochondria and chloroplasts
Cytoskeleton	Yes (but comprising different proteins to the eukaryotic cytoskeleton)	Yes
Ribosomes	Smaller (70S)	Larger (80S)
Cilia and flagella	Flagella are sometimes present, but with a different structure to eukaryotes	Often present in animal cells
Cell wall	Yes, made of peptidoglycan (murein), a polymer of sugars and amino acids	Present in plant cells (made of cellulose) and fungi (made of chitin)
Reproduction	Asexual (binary fission)	Asexual or sexual

 Go further: How do prokaryotes cope without membrane-bound organelles?

You might be wondering how bacteria carry out aerobic respiration without mitochondria. Approximately 50 years ago, infoldings of bacterial plasma membranes were observed under electron microscopes. These were called mesosomes and proposed as the prokaryotic equivalent of mitochondrial cristae. They are now thought to be artefacts (not occurring naturally in bacterial cells).

1 Suggest how the infoldings of bacterial plasma membranes may have been produced.

2 Suggest how aerobic respiration can occur in bacteria.

Chapter 2 Practice questions

1   The actual diameter of an *Amoeba proteus* cell is 0.25 mm. The cell is studied under a light microscope using a magnification of ×1600.

Which of the following values represents the image radius of the cell as seen under the microscope?

A 0.4 m

B 400 cm

C 200 mm

D 20 cm   *(1 mark)*

2   Which of the following statements is / are true of the procedure you would use to differentiate between Gram-positive and Gram-negative bacteria?

1   A primary stain is applied to heat-fixed bacteria.

2   Ethanol can be added to decolourise Gram-negative bacteria.

3   A counterstain such as safranin is applied.

A 1, 2, and 3 are correct

B Only 1 and 2 are correct

C Only 2 and 3 are correct

D Only 1 is correct   *(1 mark)*

3   Which of the following statements is true of a transmission electron microscope?

A A maximum resolution of 0.5 nm can be achieved.

B A maximum magnification of ×100 000 can be achieved.

C 3D images can be produced.

D The preparation of samples is non-invasive.   *(1 mark)*

4   Complete the following table by writing YES or NO in the blank boxes.

Structure	Membrane-bound	Contains DNA	
Mitochondrion			*(1 mark)*
Chloroplast			*(1 mark)*
Lysosome			*(1 mark)*
Microtubules			*(1 mark)*

5   The following passage describes the preparation of samples for examination under a light microscope. Complete the passage by choosing the most appropriate word to place in each gap.

Samples can be preserved by .... with chemicals such as formaldehyde. .... staining enables tissues and organelles within a sample to be distinguished. For example, Gram staining involves the application of .... to identify bacteria based on the structure of their ....   *(4 marks)*

6   The cell body of a motor neuron had a diameter of three eyepiece units. What is the diameter of the cell body in standard form?   *(2 marks)*

stage scale: 1 division = 10 µm

0   10   20

eyepiece scale

7   Compare the structure and function of flagella in eukaryotic and prokaryotic cells.   *(3 marks)*

3.1 Biological elements

Specification reference: 2.1.2(b), (c), and (p)

Revision tip: The long and the short of it

Polymers are long chains of smaller molecules (**monomers**) that have been bonded together. We will discuss examples of monomers and polymers in subsequent topics.

Revision tip: Organic learning

Organic substances contain carbon bonded to hydrogen; inorganic substances lack C–H bonds. Hydrogen carbonate (HCO_3^-) is considered inorganic because its carbon atom is bonded only to oxygen.

In this chapter you will learn about a range of important biological molecules. First you will look at the chemistry of the molecules you will be studying. You will also learn about some of the common ions found in biological systems.

The elements in biological molecules

The following table shows the elements that are always present in biological molecules or sometimes present.

Element	Carbon (C)	Hydrogen (H)	Oxygen (O)	Nitrogen (N)	Phosphorus (P)	Sulfur (S)
Atomic (proton) number	6	1	8	7	15	16
Typical number of bonds formed	4	1	2	3	3–5	2
Carbohydrates (Topic 3.3)						
Lipids (Topic 3.5)						
Proteins (Topic 3.6)						
Nucleic acids (Topic 3.8)						

Common misconception: Molecules and compounds

An element is a substance composed of one type of atom (e.g. pure carbon consists only of atoms with six protons and six electrons). Molecules contain more than one atom and can be either elements (e.g. O_2 or N_2) or compounds (which comprise more than one element, for example glucose – $C_6H_{12}O_6$). Compounds contain either ionic bonds between elements (e.g. NaCl) or covalent bonds (e.g. $C_6H_{12}O_6$). Ionic compounds are *not* molecules. They have giant lattice structures.

Inorganic ions

Ions are atoms or molecules that have lost electrons (positively charged ions – cations) or gained electrons (negatively charged ions – anions). Several inorganic ions play essential roles in organisms.

▼ **Table 1** *Inorganic ions in biology*

Ion	Examples of roles	Topic reference
Ca^{2+} (calcium)	Nervous impulse transmission	Action potentials (Year 2)
Na^+ (sodium)		
K^+ (potassium)		
H^+ (hydrogen)	Determination of pH in solutions	4.2, Factors affecting enzyme activity
NH_4^+ (ammonium)	Sources of nitrogen for plants	The nitrogen cycle (Year 2)
NO^{3-} (nitrate)		
HCO_3^- (hydrogen carbonate)	Transport of respiratory gases	8.4, Transport of oxygen and carbon dioxide in the blood
Cl^- (chloride)		
PO_4^{3-} (phosphate)	Nucleic acid and ATP formation	3.8, Nucleic acids, and 3.11, ATP
OH^- (hydroxide)	Determination of pH in solutions	4.2, Factors affecting enzyme activity

Summary questions

1   State the three elements found in carbohydrates, proteins, lipids, and nucleic acids.   (*3 marks*)

2   Explain the difference between a compound and a molecule.   (*3 marks*)

3   State which of the ions in the list below are
    a inorganic ions **b** organic ions **c** compound ions
    Cl^-, HCO_3^-, NH_4^+, CH_3COO^-, Ca^{2+}, OH^- (*3 marks*)

3.2 Water

Specification reference: 2.1.2(a)

Molecules interact with each other in a variety of ways, depending on the atoms they contain. Water molecules form hydrogen bonds with each other, which gives water the properties that make it crucial for life.

Hydrogen bonding

Water molecules (H_2O) are polar (i.e. the oxygen atom has a partial negative charge and the hydrogen atoms have partial positive charges). As a consequence, the oxygen atom in a water molecule is attracted to hydrogen atoms in neighbouring molecules. This attraction is called a **hydrogen bond**.

Properties of water

Hydrogen bonds are strong in comparison to other intermolecular forces. The unique properties of water are a result of hydrogen bonding. These properties enable water to play a pivotal role in biology.

▼ **Table 1** The properties of water

Property	Explanation	Importance
Good solvent	Polar water molecules attract (and dissolve) other polar molecules and ions	Water transports dissolved solutes (e.g. in blood and phloem) Chemical reactions occur in water
High specific heat capacity	A relatively large amount of energy is required to increase water temperature	Thermal stability in aquatic environments and inside organisms
High heat of vaporisation	Additional energy is needed to change water from liquid to gas	Thermoregulation – sweating and panting can cool an organism when water on the body is evaporated
Cohesion	Hydrogen bonds cause water molecules to be attracted to each other and flow together	Water movement up xylem vessels Surface tension – small organisms can move on the water surface
Low density solid (i.e. ice)	The crystalline structure in ice is less dense than liquid water	Provides an insulating layer for aquatic habitats in cold climates The ice surface provides a habitat for some organisms (e.g. polar bears)

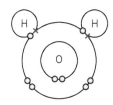

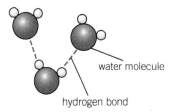

tends to pull the electrons slightly in this direction

small positive charge δ^+ on each hydrogen atom

corresponding negative charge δ^- on oxygen atom

▲ **Figure 1** Water is a polar molecule

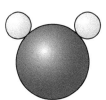

water molecule

hydrogen bond

▲ **Figure 2** Hydrogen bonds between water molecules

Synoptic link

You will study water cohesion and adhesion in xylem vessels in Topic 9.3, Transpiration, and its role as a solvent in Topic 8.3, Blood, tissue fluid, and lymph.

Summary questions

1 Explain why water is a polar molecule.   *(3 marks)*

2 Explain how the properties of water are suited to its role as a transport medium in blood and xylem vessels.   *(3 marks)*

3 Amino acids exist as zwitterions in solution (i.e. each amino acid molecule has both positive and negative charges).
   Explain why zwitterions dissolve in water.   *(2 marks)*

3.3 Carbohydrates

Specification reference: 2.1.2(d), (e), (f), and (g)

Carbohydrates are molecules containing only carbon, oxygen, and hydrogen atoms. The chemical energy stored in the bonds of some carbohydrates can be used by organisms. Other carbohydrates have structural and storage roles. Here you will learn about a range of carbohydrates vital to biology.

Monosaccharides

A monosaccharide is the simplest carbohydrate unit (monomer). Two examples of monosaccharides that you will encounter in future topics are **glucose** and **ribose**.

▼ **Table 1** *The structures and functions of glucose and ribose*

Monosaccharide	Number of carbon atoms	Formula	Structure	Use
Glucose	6 (hexose sugar)	$C_6H_{12}O_6$	α-glucose β-glucose	**α-glucose:** a substrate in respiration (a Year 2 topic) **β-glucose:** polymerises to form cellulose (see 'Polysaccharides')
Ribose	5 (pentose sugar)	$C_5H_{10}O_5$		The sugar in RNA nucleotides (Topic 3.8, Nucleic acids) and ATP (Topic 3.11, ATP)

Disaccharides

Two monosaccharides can react together to form a disaccharide. This reaction is called a **condensation reaction**; it creates a **glycosidic bond** between the two monosaccharides and produces water. Disaccharides can be broken down to reform the original two monosaccharides in a **hydrolysis** reaction.

▲ **Figure 1** *The formation and breakdown of lactose*

▼ **Table 2** *The three disaccharide sugars and their constituent monomers*

Monosaccharides	Disaccharide produced
Glucose + glucose	Maltose
Glucose + galactose	Lactose
Glucose + fructose	Sucrose

Polysaccharides

Polysaccharides are long carbohydrate molecules (polymers) formed when many monosaccharides bond together in condensation reactions. Four examples of polysaccharides formed from glucose monosaccharides are outlined in Table 3.

▼ **Table 3** *The polysaccharides made from glucose monomers*

	Glycogen	Starch		Cellulose
		Amylose	**Amylopectin**	
Monomer	α-glucose	α-glucose		β-glucose
Type of glycosidic bonds	1,4 and 1,6 links	1,4 links	1,4 and 1,6 links	1,4 links
Branching	Yes	No	Yes	No
Helical	Yes	Yes	No	No
Function	Carbohydrate storage in animals	Carbohydrate storage in plants		Structural support in plant cell walls
Properties that suit function	Insoluble. Compact due to branching. The number of points at which glucose can be released through hydrolysis is increased by branching.	Amylose is insoluble. Helices and branching make starch compact. Branching in amylopectin increases the number of points at which glucose can be released through hydrolysis.		Insoluble. Cross links (hydrogen bonds between chains) increase strength.

Revision tip: Sugars, short and sweet

Polysaccharides are not classed as sugars, which are short-chain carbohydrates (i.e. monosaccharides and disaccharides).

Summary questions

1   Describe how sucrose is formed.   *(3 marks)*

2   Describe two similarities and one difference between the structures of glucose and ribose.   *(3 marks)*

3   Contrast the structures of cellulose and glycogen. Explain how the two structures are suited to the functions of the polysaccharides.   *(6 marks)*

3.4 Testing for carbohydrates

Specification reference: 2.1.2(q) and (r)

You have already learned about a variety of carbohydrates. Now you will examine how to test for the presence of these carbohydrates.

Practical skill: Chemical tests

▼ **Table 1** *Chemical tests for starch, reducing sugars and non-reducing sugars*

	Benedict's test		Iodine test
Carbohydrate being identified	Reducing sugars (e.g. monosaccharides, lactose, maltose)	Non-reducing sugars (e.g. sucrose)	Starch
Description	Mix with Benedict's reagent in a boiling tube and heat	After a negative result with the Benedict's test, boil with dilute HCl. Conduct the Benedict's test a second time.	Mix iodine/ potassium iodide solution with the sample
Negative result	**BLUE**		**YELLOW/BROWN**
Positive result (i.e. the carbohydrate is present)	**LOW = GREEN** **MEDIUM = ORANGE** **HIGH = RED**		**PURPLE/BLACK**

Revision tip: Reagent strips

An alternative to the Benedict's test and colorimetry is the use of reagent strips, which can be used to test for reducing sugars. The concentration of sugar is assessed by comparing the strip colour to a chart.

Colorimetry

A colorimeter can be used to assess the concentration of sugar in a solution.

Practical skill: Using a colorimeter following the Benedict's test

1. Insert a **red filter**.
2. Use a cuvette of distilled water to zero (**calibrate**) the colorimeter.
3. Construct a **calibration curve** (i.e. perform the Benedict's test on a series of solutions with known glucose concentrations, filter the precipitate, obtain a colorimeter reading (either transmission or absorption) for each solution, and plot a graph of transmission (or absorption) against concentration).
4. Conduct the Benedict's test on your test solution, then **filter the precipitate**.
5. Obtain a colorimeter reading for the solution and calculate the glucose concentration from the calibration curve.

Practical skill: Biosensors

Biosensors represent another method for determining the concentration of sugars in a solution. These machines require a recognition molecule that will bind the carbohydrate being assessed. The extent to which the carbohydrate binds determines the reading on the biosensor display. Unlike the Benedict's test, biosensors detect specific sugars (e.g. glucose).

Summary questions

1. State three quantitative methods for determining the concentration of glucose in a solution.
   (3 marks)

2. Sketch a colorimeter calibration curve for glucose concentration against **a** absorption **b** transmission. Assume that the precipitate is re-moved after the Benedict's test. Explain the shape of the absorption graph.
   (6 marks)

3. In colorimetry, a red filter is used when measuring transmission through solutions of Benedict's reagent. Explain why.
   (2 marks)

3.5 Lipids

Specification reference: : 2.1.2(h), (i), (j), and (q)

Like carbohydrates, lipids contain carbon, oxygen, and hydrogen. Unlike many carbohydrates, lipids are non-polar and largely insoluble in water, which means the two types of molecule occupy different roles in organisms.

Examples of lipids

Lipids vary widely in structure. Three examples are triglycerides, phospholipids, and cholesterol.

Synoptic link

In Topic 5.1, The structure and function of membranes, you will learn more about the roles of phospholipids and cholesterol in membranes.

▼ **Table 1** *Structures and functions of three lipids*

	Triglyceride	Phospholipid	Cholesterol
Structure	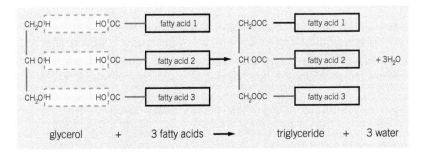		
Properties	Compact and insoluble	Hydrophilic head and hydrophobic tail	Small; both hydrophilic and hydrophobic
Roles	Energy storage and insulation	Membrane structure	Membrane stability and steroid hormones

Synthesis of triglycerides

A triglyceride comprises a **glycerol** molecule and **three fatty acids**. Each fatty acid undergoes a **condensation reaction** with one of the OH (alcohol) groups in glycerol. A **hydrolysis** reaction breaks down the triglyceride into the original glycerol and fatty acids.

▲ **Figure 1** *Triglyceride synthesis via condensation reactions*

Revision tip: Saturated vs unsaturated

Fatty acids can be either saturated (no C=C double bonds) or unsaturated (at least one C=C double bond). Evidence exists that excessive consumption of saturated triglycerides raises the risk of coronary heart disease.

Practical skill: The emulsion test

Lipids are identified using the emulsion test.

1 Mix your sample with **ethanol**
2 Mix with **water** and **shake**
3 The formation of a **white emulsion** on top of the solution indicates the presence of a lipid.

Summary questions

1 State what would be seen when an emulsion test gives a positive result. *(2 marks)*

2 Explain why the structure of a phospholipid is suited for its role in membranes. *(4 marks)*

3 Outline the similarities and differences between the synthesis of polypeptides and the synthesis of triglycerides. *(4 marks)*

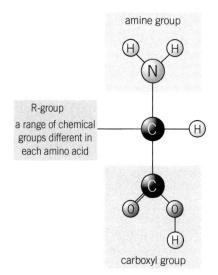

amine group

R-group
a range of chemical groups different in each amino acid

carboxyl group

▲ **Figure 1** *The structure of an amino acid*

Synoptic link

You will learn more about polypeptide formation during translation in Topic 3.10, Protein synthesis.

Practical skill: The biuret test

Proteins are identified using the biuret test. This is a relatively simple test that does not require heat.

1 Add sodium hydroxide (NaOH) to the sample (equal volumes).

2 Add drops of copper sulphate solution (which is blue) and mix.

3 A purple colour shows the presence of protein.

Revision tip: Primary structure determines secondary structure

A protein's primary structure (the sequence of amino acids in a polypeptide) determines how the polypeptide folds or coils. The arrangement of R groups in the amino acid sequence dictates which bonds form and their locations. This influences both secondary and tertiary structure.

Proteins consist of amino acid monomers bonded together in chains. Here you will learn about the structure of amino acids. We also discuss how the secondary, tertiary, and quaternary structures of proteins produce their specific shapes.

Amino acids

All amino acids contain a central carbon atom, an amine group, a carboxyl group, and a hydrogen atom. The identity of an amino acid is determined by its R group; this is different in each of the 20 amino acids found in organisms.

Practical skill: Thin layer chromatography (TLC)

What does TLC do? It separates particles in a mixture (e.g. a solution of different amino acids).

How does TLC work? The mixture is applied at the base (the *origin*) of a silica gel layer (*stationary phase*), which is dipped into an organic solvent (*mobile phase*). Amino acids dissolve in the solvent as it diffuses up the gel. The rate at which the amino acids move up the gel depends on how much they interact (*adsorb*) with the gel. Amino acids will interact to different extents and will therefore move different distances.

How can the amino acids be visualised? The position of the amino acids can be seen by spraying ninhydrin onto the gel. This reveals each set of amino acids as a purple/brown spot.

How are the amino acids identified? The identity of each amino acid is revealed by calculating R_f values. Each amino acid has a unique R_f value under particular conditions.

$$R_f = \frac{\text{distance moved by the amino acid}}{\text{distance moved by the solvent}}$$

Polypeptides

Amino acid monomers are joined together by **condensation reactions**. A **peptide bond** is formed between two amino acids, producing a **dipeptide**. A **polypeptide** is formed when many amino acids bond together. This occurs at ribosomes during **translation**. Peptide bonds are broken in hydrolysis reactions.

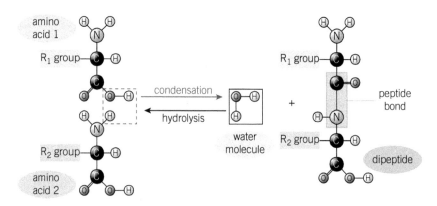

▲ **Figure 2** *Dipeptide synthesis and breakdown*

Levels of protein structure

A polypeptide produced during translation is called the **primary structure** of a protein. However, proteins are more complex than a sequence of amino acids in a chain. They fold and coil into shapes specific to their functions; these are called **secondary** and **tertiary structures**.

	Secondary structure	Tertiary structure
Bonds involved	Hydrogen bonds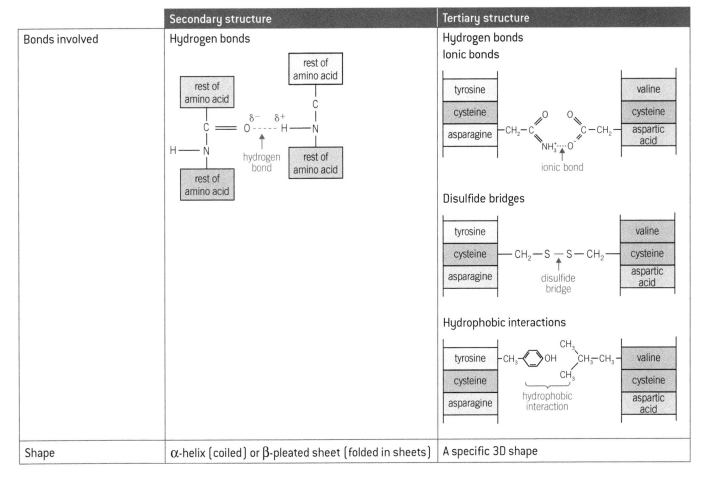	Hydrogen bonds Ionic bonds Disulfide bridges Hydrophobic interactions
Shape	α-helix (coiled) or β-pleated sheet (folded in sheets)	A specific 3D shape

Quaternary structures are produced when two or more polypeptides associate. For example, haemoglobin consists of four polypeptide subunits and collagen comprises three polypeptides. Prosthetic groups (see Topic 4.4, Cofactors, coenzymes, and prosthetic groups) can be present in quaternary structures.

Common misconception: Polypeptide vs protein

A protein is more than a chain of amino acids. The polypeptide that is translated at ribosomes is a linear sequence of amino acids (the primary structure). Proteins, however, exhibit specific, complex shapes (due to bonds in their secondary and tertiary structures) and can contain several polypeptides.

 Go further: Gluten

Fans of *The Great British Bake Off* hear gluten mentioned on a regular basis. Gluten is a protein that is pivotal to baking bread. Except ... it's not just one protein. Gluten is a mixture of two types of protein: gliadins (which are smaller) and glutenins (which are larger).

Dough is kneaded to stretch gluten molecules. This allows more intermolecular bonds (e.g. hydrogen bonds) to form between glutenin and gliadin polypeptides, making the dough stronger and trapping CO_2 under gluten sheets.

1 Gluten proteins are insoluble in water. What does this suggest about the nature of the majority of their amino acids?

2 Explain why too much kneading of dough can create chewy bread.

3 Salt is added to dough to strengthen it by allowing gluten molecules to move closer to each other. Suggest how salt enables glutenins and gliadins to move closer together.

Summary questions

1 Draw alanine, which is an amino acid that has CH_3 as its R group. (*2 marks*)

2 Two amino acids were analysed using thin layer chromatography. The solvent was allowed to travel 5 cm along the gel. Calculate **a** the R_f value of amino acid X, which travelled 4.15 cm. **b** the distance travelled by amino acid Y, which has an R_f value of 0.71. (*2 marks*)

3 Suggest how the biuret test can be used to assess the concentration of proteins. (*4 marks*)

3.7 Types of proteins

Specification reference: 2.1.2(n) and (o)

Synoptic link

We will explore the role of haemoglobin further in Topic 8.4, Transport of oxygen and carbon dioxide in the blood.

Synoptic link

You will learn more about prosthetic groups in Topic 4.4, Cofactors, coenzymes, and prosthetic groups.

Revision tip: Conjugated proteins

Some globular proteins are conjugated, which means they contain a non-protein component called a prosthetic group. Prosthetic groups can be metal ions, lipids, carbohydrates, or molecules derived from vitamins.

In Topic 3.6, Structure of proteins, you learned how the tertiary and quaternary structures of proteins are produced. Proteins can be divided into two broad categories, based on the nature of these structures: **globular** and **fibrous** proteins. Here we compare the structures and properties of these two groups of proteins.

Globular and fibrous proteins

▼ **Table 1** *A comparison of globular and fibrous proteins*

	Globular	Fibrous
Shape	Compact and spherical	Long and linear
Bonding/structure	Hydrophilic R groups on the outside, and hydrophobic R groups on the inside	A limited range of amino acids, often with a repetitive sequence Organised and strong structures
Water solubility	Soluble	Insoluble
Conjugation (i.e. presence of prosthetic group)	Sometimes	No
Functions	Enzymes (e.g. catalase) Hormones (e.g. insulin) Membrane proteins Antibodies Transport proteins (e.g. haemoglobin, which has four polypeptides in its quaternary structure, each carrying a haem prosthetic group)	Structural roles For example: Keratin (in skin, nails, and hair) Collagen (in connective tissue in tendons, skin, and ligaments)

Summary questions

1 In which organelle are prosthetic groups added to polypeptides?

(1 mark)

2 Explain, in terms of their functions, why it is important for globular proteins to be soluble and for fibrous proteins to be insoluble.

(2 marks)

3 Keratin contains many disulfide bonds. Keratin in hair contains fewer disulfide bonds than keratin in nails. Suggest why. *(2 marks)*

3.8 Nucleic acids

Specification reference: 2.1.3(a), (b), and (d)(i), and (ii)

You have learned about two types of polymer (polysaccharides and polypeptides) in earlier topics. Here we introduce another class of polymer, nucleic acids, which are built from nucleotide monomers.

Nucleotide structure

A nucleotide has three components: a pentose (5-carbon) sugar, a phosphate group, and an organic nitrogenous base.

Nucleic acid polymers

Deoxyribonucleic acid (DNA) and **ribonucleic acid (RNA)** are both polynucleotides. You will discover more about their roles in the next couple of topics. Polynucleotides are formed by **condensation reactions** between nucleotides, which form **phosphodiester bonds**. As with other polymers you have encountered, polynucleotides are broken by **hydrolysis**.

DNA

The sugar in DNA nucleotides is deoxyribose. Each DNA nucleotide has one of four bases: adenine (A), thymine (T), cytosine (C), or guanine (G).

Complementary base pairing

DNA exists as a double helix (i.e. two polynucleotides, running in opposite directions, bonded together). The two DNA polynucleotide strands are held together by hydrogen bonds between their bases. C on one strand bonds with G on the other strand; similarly, A always bonds with T.

▼ **Table 1** *Complementary base pairs in DNA*

Purine base (double-ring structure)	Number of hydrogen bonds between the bases	Pyrimidine base (single-ring structure)
Adenine	Two hydrogen bonds between A and T	Thymine
Guanine	Three hydrogen bonds between G and C	Cytosine

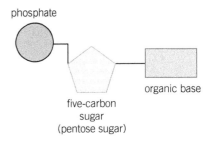
▲ **Figure 1** *The generalised structure of a nucleotide*

Synoptic link

In Topic 3.11, ATP, we will discuss the importance of ATP as a phosphorylated nucleotide.

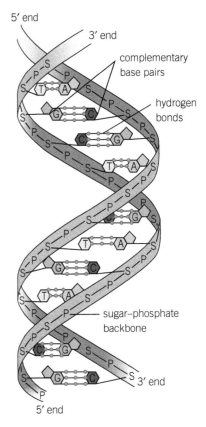
▲ **Figure 2** *DNA double helix – complementary base pairing between antiparallel strands*

Revision tip: Basic memory tricks

A mnemonic is a memory tool. Mnemonics can sharpen memories by making them more visual. For example, *small pyramid CiTies; big, pure GArdens* may help you remember the relative size of DNA bases, and whether they are purines or pyrimidines.

Synoptic link

You learned about the structure of ribose in Topic 3.3, Carbohydrates.

RNA

RNA exists in three forms (messenger, transfer, and ribosomal RNA), all of which play roles in polypeptide synthesis. You will learn more about RNA in Topic 3.10, Protein synthesis. The structure of RNA is different to DNA in the following ways:

- The pentose sugar in RNA is **ribose** (rather than the deoxyribose in DNA).
- **Uracil** (U) is present instead of thymine; U bonds with A in RNA.
- RNA is **single-stranded** and does *not* form a double helix.

Practical skill: DNA purification

You can use the following steps to extract DNA from plant tissue:

1 **Grind** tissue (to break down cell walls)
2 Mix with **detergent** (to break down membranes)
3 Add **salt** (to break hydrogen bonds between DNA polynucleotides)
4 Add **protease** enzyme (to break down proteins surrounding DNA)
5 Add **alcohol** (to precipitate DNA out of solution)
6 Remove DNA (which can be seen as white strands below the alcohol layer).

Summary questions

1 Complete the table below, which features monomers and polymers you have encountered in this chapter. *(3 marks)*

Monomer	Polymer	Bond formed in condensation reaction
Monosaccharide		
	Polypeptide	
Nucleotide		

2 Explain why the DNA of a cell always contains equal amounts of adenine and thymine, and equal amounts of cytosine and guanine. *(2 marks)*

3 Suggest why C–A and T–G base pairings do not occur. *(2 marks)*

3.9 DNA replication and the genetic code

Specification reference: 2.1.3(e) and (f)

In Topic 3.8, Nucleic acids, you looked at the structure of DNA. Genetic material must be copied whenever a cell divides. In this topic we examine how DNA replication is achieved. You will also learn that the genetic code is degenerate and works through sequences of triplets.

DNA replication

The following stages occur in DNA replication:

1 Histone proteins are removed

2 The DNA double helix is **unwound** (by an enzyme called *helicase*)

3 Hydrogen bonds between strands are broken (**unzipping**; catalysed by *helicase*)

4 Both DNA strands act as **template strands**

5 Free (monomer) nucleotides are activated (phosphate groups are added to them)

6 Free nucleotides form hydrogen bonds with bases on the template strands

7 C bonds with G, A bonds with T (**complementary base pairing**)

8 Phosphodiester bonds join nucleotides in the new strands (catalysed by *DNA polymerase*).

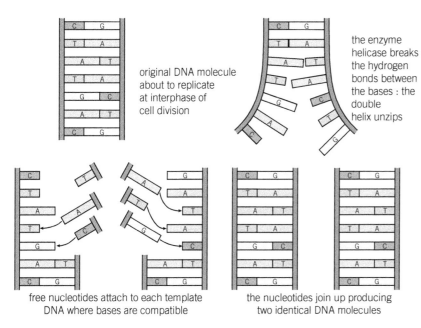

original DNA molecule about to replicate at interphase of cell division

the enzyme helicase breaks the hydrogen bonds between the bases : the double helix unzips

free nucleotides attach to each template DNA where bases are compatible

the nucleotides join up producing two identical DNA molecules

▲ **Figure 1** *Semi-conservative DNA replication*

The genetic code

The genetic code has the following properties:

• It uses a **triplet code** (i.e. a sequence of three nucleotides, known as a **codon**, codes for the production of one amino acid).

• The code is **degenerate** (i.e. in most cases, several codons code for the same amino acid). For example, TCT, TCC, TCA, TCG, AGT, and AGC all code for serine.

Summary questions

1 The base sequence GCCAAATCT appears in one strand of DNA. State the bases that would be present in the opposite strand. *(3 marks)*

2 Outline the importance of enzymes in DNA replication. *(4 marks)*

3 Explain why a genetic mutation (a change in the DNA sequence of a gene) does not always result in a different protein being produced. *(2 marks)*

3.10 Protein synthesis

Specification reference: 2.1.3(g)

Key terms

Transcription: The production of an mRNA molecule from a DNA base sequence (i.e. a gene).

Translation: The production of a polypeptide at ribosomes; the sequence of amino acids is determined by the codons in mRNA.

Synoptic link

Topic 2.4, Eukaryotic cell structure, provided an outline of the organelles involved in protein synthesis.

Revision tip: Replication vs transcription

The start of DNA replication and transcription are similar; DNA is unwound and hydrogen bonds between complementary bases are broken. However, the two processes show many differences. For example, during transcription only a small section of DNA is unzipped, only one strand acts as a template, RNA polymerase catalyses the reaction rather than DNA polymerase, and mRNA is produced rather than a new DNA strand.

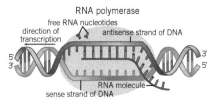

▲ **Figure 1** *Transcription: the role of RNA polymerase*

Two processes are needed to convert a genetic blueprint into a polypeptide: **transcription** and **translation**.

Transcription

Transcription produces a messenger RNA (mRNA) strand from a DNA sequence (a gene) using these steps:

1 *Helicase* **unwinds** and **unzips** DNA (but usually only along the base sequence of one gene).
2 **RNA nucleotides** bind to complementary DNA bases on the **template (antisense)** strand.
3 *RNA polymerase* joins RNA nucleotides together with phosphodiester bonds.
4 A **stop codon** causes RNA polymerase to detach.
5 The mRNA molecule detaches from the DNA and leaves the nucleus.

Translation

Translation occurs at ribosomes and involves the following stages:

1 **mRNA attaches** to a ribosome (in a groove between the two subunits).
2 The first mRNA **codon** binds to a **tRNA** molecule with a complementary **anticodon**.
3 The tRNA is bound to an amino acid specific to its anticodon.
4 A second tRNA (carrying another amino acid) binds to the adjacent mRNA codon.
5 A **peptide bond** forms between the two amino acids.
6 The ribosome continues along the mRNA molecule, from codon to codon.
7 Amino acids are bonded together in a polypeptide chain until a **stop codon** is reached.
8 The polypeptide is released.

Summary questions

1 Which DNA strand acts as the template strand during transcription? *(1 mark)*

2 Explain why DNA must be double-stranded despite only one strand representing the genetic code. *(2 marks)*

3 Complete Table 1, which compares DNA replication, transcription, and translation. *(7 marks)*

▼ **Table 1**

	DNA replication	Transcription	Translation
Hydrogen bonds between complementary DNA base pairs are broken			
Free DNA nucleotides are activated			
To which base does adenine bond?			
Phosphodiester bonds are formed			
Peptide bonds are formed			
Location			
Product			

3.11 ATP

Specification reference: 2.1.2(b) and 2.1.3(c)

You were introduced to the structure of nucleotides in Topic 3.8, Nucleic acids. Adenosine triphosphate (ATP) is a phosphorylated nucleotide. It is the currency of chemical energy found in all organisms. In this topic you will learn about the structure of ATP and why it is suited to energy transfer.

ATP structure

ATP is similar in structure to an RNA nucleotide. It contains an adenine base and a ribose sugar. ATP, however, has three phosphate groups rather than the one found in an RNA nucleotide.

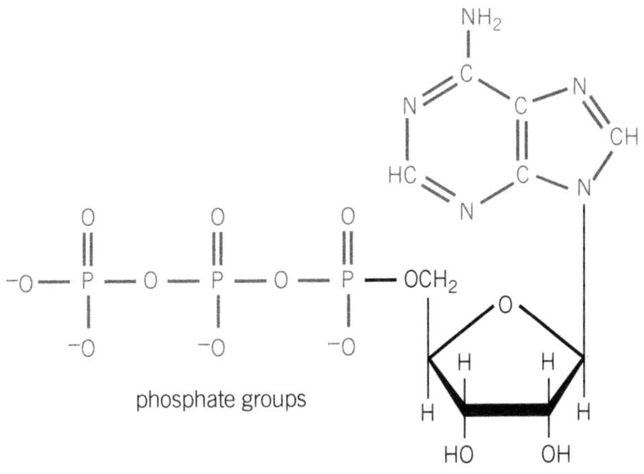

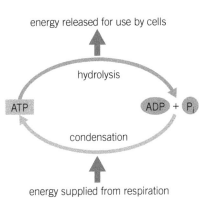

▲ **Figure 1** *ATP structure*

> **Revision tip: How does ATP deliver energy?**
> ATP releases energy to enable a vast range of reactions to occur. How is the energy transferred? A relatively small amount of energy is needed to remove a phosphate from ATP. Much more energy is released when the phosphate forms a new bond with another substance. Overall, energy is liberated and can be used in other metabolic processes.

Condensation and hydrolysis

ATP is formed in respiration through a condensation reaction. Hydrolysis of ATP produces ADP by removing one of the phosphates. This hydrolysis reaction releases energy that can be used in metabolic processes.

▲ **Figure 2** *The cycle of ATP production and use*

How is ATP suited to energy transfer?

The structure of ATP makes it an excellent energy carrier because each molecule:

- is soluble
- releases energy in small quantities (which prevents energy wastage)
- has an unstable phosphate bond (which is easily broken)

Summary questions

1 Outline the similarities and differences between RNA nucleotides and ATP. *(4 marks)*

2 Explain why ATP is not a suitable long-term energy store. *(2 marks)*

3 ATP is described as a universal energy carrier. Suggest the meaning of the term 'universal energy carrier'. *(2 marks)*

1 Which of the following structural properties is possessed by glycogen?

   A Contains only 1,4 glycosidic bonds

   B Branched

   C Formed from β-glucose

   D Formed from monomers with the formula $C_5H_{10}O_5$ *(1 mark)*

2 Which of the following statements is a correct step in the emulsion test for lipids?

   A Mix the sample with HCl

   B Mix the sample with ethanoic acid

   C Heat the mixture

   D Shake the mixture   *(1 mark)*

3 Which of the following bonds is the weakest of the bonds that hold the tertiary structure of proteins together?

   A Hydrogen   **C** Ionic

   B Disulfide   **D** Peptide   *(1 mark)*

4 Which of the following statements is true of fibrous proteins?

   A They can be conjugated

   B Their amino acid sequences can be repetitive

   C They are soluble

   D Antibodies are examples of fibrous proteins   *(1 mark)*

5 The following passage describes the construction of a calibration curve when using colorimetry to measure the concentration of fructose. Complete the passage by choosing the most appropriate word to place in each gap.

A .... filter is placed in the colorimeter, which is calibrated using a .... of distilled water. A calibration curve is obtained by performing the .... test on solutions with known fructose concentrations. The .... is filtered and colorimeter readings are obtained for each solution.   *(4 marks)*

6 **a** Phosphorus is an element that is essential for plant growth. Describe how phosphorus is absorbed by plants.   *(3 marks)*

   b Outline the biological roles of phosphorus.   *(6 marks)*

7 **a** Explain the role of water in temperature regulation in organisms.

       (4 marks)

   b Draw a diagram to show hydrogen bonding between a water molecule and an ethanol (CH_3CH_2OH) molecule.   *(2 marks)*

8 **a** Explain the purpose of adding the following substances in the extraction of DNA from plant material.   *(4 marks)*

      • detergent   • salt   • protease   • alcohol

   b These questions concern the DNA base sequence ACGTTA.

      i Describe the sequence in terms of purines and pyrimidines. *(2 marks)*

      ii State the number of amino acids coded for by this sequence.   *(1 mark)*

      iii State the mRNA base sequence that would be transcribed by this DNA base sequence.   *(2 marks)*

Many globular proteins are enzymes, which are biological catalysts.

The importance of enzymes

Metabolism is the term given to the sum of all the chemical reactions occurring in an organism. Enzymes control these metabolic reactions, which can either be **anabolic** (the formation of molecules from smaller units) or **catabolic** (the breaking down of molecules).

An example of an anabolic reaction is DNA replication, which is controlled by the enzyme DNA polymerase (see Topic 3.9, DNA replication and the genetic code). Digestion is a catabolic process; for example, the enzyme maltase breaks down maltose into two molecules of glucose.

Enzymes can be either intracellular (working inside cells) or extracellular (working outside cells). Most extracellular enzymes are catabolic. In multicellular organisms, digestive enzymes are secreted from cells into the digestive system. Unicellular organisms, such as yeast and bacteria, secrete digestive enzymes into their immediate environment.

How do enzymes work?

All enzymes operate through the same principles:

1. **Substrates** collide with the **active site** of the enzyme.
2. The shape of the active site is **complementary** to the substrate.
3. The substrate binds to the active site to form an **enzyme–substrate complex** (ESC).
4. Bonds in the substrates are placed under strain and break; the enzyme provides an alternative reaction pathway that reduces the **activation energy** required for the reaction.
5. An **enzyme–product complex** is formed and the product(s) is/are released.

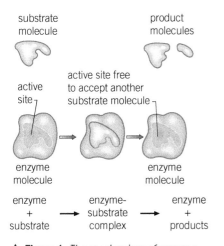

▲ **Figure 1** The mechanism of enzyme action (illustrated here by a catabolic reaction)

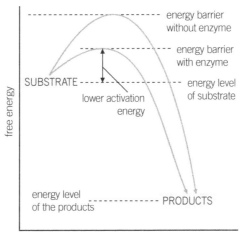

▲ **Figure 2** Enzymes lower the activation energy of a reaction (i.e. less energy is required to initiate the reaction)

Two competing theories exist to explain the enzyme–substrate binding mechanism. The **lock and key** hypothesis suggests that the shape of the active site is an ideal fit for the substrate molecule and is therefore **specific** to one substrate. The **induced-fit** hypothesis, however, suggests that initially weak binding by the substrate will alter the enzyme's tertiary structure. This strengthens the temporary bonds between the substrate and the enzyme, and weakens bonds within the substrate.

Key term

Enzymes: Biological catalysts that facilitate chemical reactions.

Synoptic link

You learned about globular proteins in Topic 3.7, Types of proteins.

Key terms

Substrate: A reactant in an enzyme-catalysed reaction.

Active site: The region of an enzyme to which substrates bind.

Revision tip: Enzyme names

Enzymes often end with the suffix '-ase' and are named after the substrate on which they act. Lactase, for example, breaks down lactose, and glycogen synthase catalyses the synthesis of glycogen. Some enzymes, however, do not conform to this naming system (e.g. trypsin breaks down polypeptides into smaller peptides).

Summary questions

1 State three structural features that are common to all enzymes. (*3 marks*)

2 State whether the following reactions are anabolic or catabolic. Explain your answers.
a The formation of glycogen b The formation of maltose from amylose
c Glycerol formation from a triglyceride (*3 marks*)

3 Explain how enzymes are able to lower the activation energy of reactions. (*4 marks*)

You learned about how enzymes facilitate biological reactions in Topic 4.1, Enzyme action. An enzyme's performance, however, is dependent on several factors, which we discuss here.

Which factors affect enzyme activity?

The following table shows the principal factors affecting the performance of enzymes.

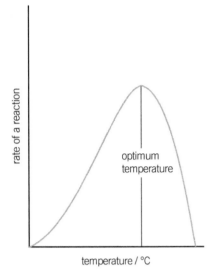

▲ **Figure 1** *All enzymes have an optimum temperature at which they work*

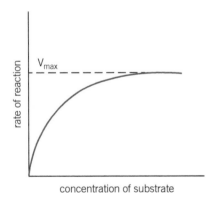

▲ **Figure 2** *Reaction rates increase as substrate concentration increases until V_{max} is reached*

Factor	Effect	Explanation
Temperature	A rise in temperature will increase enzyme activity up to an optimum temperature	Enzyme and substrate molecules gain kinetic energy and move faster, thereby increasing the chance of successful collisions
	Above the optimum temperature, further increases in temperature reduce and eventually stop enzyme activity	Weak bonds (e.g. hydrogen bonds) in the active site vibrate more, strain, and break (i.e. the enzyme **denatures**)
pH	Enzyme activity decreases when pH moves away from the optimum (either decreasing or increasing)	Changes to H^+ ion concentrations alter the charges on amino acids in the active site, which can prevent substrate molecules from binding
	A further change in pH can stop all enzyme activity	Hydrogen and ionic bonds in the active site are broken, which causes a permanent change in the enzyme's tertiary structure (i.e. the enzyme is denatured)
Substrate concentration	Reaction rate increases as substrate concentration rises but eventually plateaus (when V_{max}, the maximum rate of reaction, is reached). Enzyme concentration becomes a limiting factor when V_{max} is reached.	A higher collision rate between substrates and active sites results in enzyme–substrate complexes (ESCs) forming at a greater rate. Reaction rate plateaus when one of the factors becomes limiting.
Enzyme concentration	An increase in enzyme concentration will raise the reaction rate to a higher V_{max} (at which point substrate concentration becomes the limiting factor)	

Revision tip: One enzyme's optimum is another enzyme's denaturation

Optimum temperatures and pH vary between enzymes, depending on where they operate. For example, enzymes in the human body tend to function best at approximately 37.5 °C, whereas thermophilic bacteria possess enzymes with much higher optimum temperatures. Digestive enzymes operating in the acidic environment of the stomach have a low optimum pH (e.g. pH 2), whereas the higher pH of the small intestine requires a higher optimum pH (e.g. pH 7).

Common misconception: What counts as denaturation?

All proteins, not only enzymes, can be denatured. A protein is denatured when bonds that hold its secondary or tertiary structure together are broken. These bonds might include hydrogen bonds, ionic bonds, and disulfide bridges. Bonds can be broken by temperature increases (above an optimum) and large shifts in pH. An alteration of secondary or tertiary structure usually results in a loss of function.

The following are not examples of denaturation:

- a small decrease in temperature (which slows enzyme activity due to a reduction of kinetic energy); exposure to very cold temperatures, however, can result in cold denaturation of enzymes, which you are not required to understand.

- enzyme inhibition (see Topic 4.3, Enzyme inhibitors)

- the breaking of peptide bonds in a protein's primary structure.

Key term

Denaturation: The (usually permanent) change in the tertiary structure of a protein.

Revision tip: The temperature coefficient

When assessing the effect of temperature on the rate of enzyme-controlled reactions, a good rule of thumb is that a 10 °C rise in temperature will double the reaction rate. In other words, Q_{10} (the temperature coefficient) is 2.

Practical skills: How to investigate factors that affect enzyme activity

When investigating enzyme activity, several factors can act as independent variables (e.g. temperature, pH, substrate concentration, enzyme concentration, and cofactor concentration – see Topic 4.4, Cofactors, coenzymes, and prosthetic groups). Remember, in a well-designed experiment only the independent variable will change. All other factors must be controlled. You will therefore be testing the effect of a single factor (the **independent variable**) on the rate of an enzyme-catalysed reaction (the **dependent variable**).

The values used in the investigation should be appropriate (i.e. within realistic ranges) if the results are to be **valid**.

Repetitions of the experiment should be conducted to improve the **reproducibility** of the results.

Example: investigating the effect of temperature on the activity of lipase

Independent variable = temperature (a water bath can be set to a range of temperatures, e.g. 0–60 °C)

Dependent variable = the rate at which lipids are broken down (for example, the conversion of triglycerides in milk to fatty acids can be monitored using a pH indicator)

Factors to control (which should be the same for each temperature tested) = the initial pH, lipase concentration, the volume of milk, the overall volume of solution.

Summary questions

1 Explain how high temperatures and changes in pH can prevent enzymes from functioning. *(5 marks)*

2 Suggest why microorganisms that inhabit alkaline lakes are unlikely to cause infections in humans. *(2 marks)*

3 Psychrophiles are organisms that inhabit cold environments where temperatures can drop below 0 °C. Suggest how the structure of psychrophiles' enzymes might differ from those found in humans. *(2 marks)*

4.3 Enzyme inhibitors

Specification reference: 2.1.4(f)

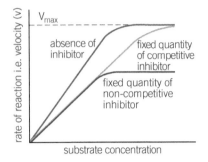

▲ **Figure 1** *The effect of substrate concentration on reaction rate in the presence and absence of inhibitors*

Common misconception: Competitive inhibitors – similar but not the same

Take care when answering questions about competitive inhibitors – the wording is important. A competitive inhibitor has a *complementary* shape to the active site of the enzyme (not the *same* shape). We can therefore conclude that the inhibitor has a *similar* shape to the substrate (but not the *same* shape, overall). The inhibitor and the substrate might, however, share some of the same groups.

Summary questions

1 State two differences in the mechanisms exhibited by competitive and non-competitive inhibitors.
(*2 marks*)

2 Sulfanilamide is an antibiotic. It inhibits a bacterial enzyme that helps to convert a molecule called PABA into folic acid. PABA and sulfanilamide have different structures, but both contain a benzene ring and an amine group. Suggest how sulfanilamide inhibits bacterial enzymes. (*4 marks*)

3 Explain why end-product inhibition is an example of negative feedback. (*2 marks*)

In Topic 4.2, Factors affecting enzyme activity, you examined how enzyme activity can be limited by factors such as temperature, pH, and the concentrations of enzyme and substrate molecules. Here we consider how certain chemicals, known as inhibitors, can bind to enzymes, thereby slowing the rate of the reactions they catalyse.

Competitive and non-competitive inhibition

Enzyme inhibitors can be either competitive or non-competitive. The following table provides a summary of the features of these two forms of inhibition.

	Competitive	Non-competitive
Binding site	Active site	Allosteric site (i.e. a region of the enzyme that is not the active site)
Mechanism	Competes with the substrate for the enzyme's active site and blocks the substrate from entering	Tertiary structure of the enzyme (and therefore active site) changes shape, meaning the active site is no longer a complementary shape to the substrate
Reversible or irreversible	Reversible (usually)	Sometimes reversible, sometimes not
Examples	*Penicillin*: an antibiotic that inhibits the active site of the bacterial enzyme transpeptidase, thereby preventing cell wall formation *Statins*: inhibit HMG-CoA reductase (an enzyme involved in cholesterol production)	*Cyanide ions (CN^-)*: a metabolic poison that inhibits cytochrome c oxidase (a respiratory enzyme) *Organophosphates*: they are used as insecticides; these chemicals inhibit acetylcholinesterase, which is an enzyme necessary for the correct functioning of insect (and mammalian) nervous systems
Effect on reaction rate	Slows rate, but V_{max} (maximum rate) can still be reached if substrate concentration is increased	Slows rate and lowers V_{max}
What does it look like?	competitive inhibitor interferes with active site of enzyme so substrate cannot bind	non-competitive inhibitor changes shape of enzyme so it cannot bind to substrate

End-product inhibition

End-product inhibition occurs at the conclusion of metabolic reaction pathways. The final product in a series of reactions will often inhibit one of the enzymes in the reaction pathway from which it has been produced. This regulates the rate at which the product is made (i.e. it acts as a negative feedback mechanism). End-product inhibition can be competitive or non-competitive.

4.4 Cofactors, coenzymes, and prosthetic groups

Specification reference: 2.1.4 (e) and (f)

In Topic 4.3, Enzyme inhibitors, you learned about substances that inhibit enzyme activity. Now you will examine substances that have the opposite effect and are required for enzymes to work. These substances are known as cofactors.

Categories of cofactors

Cofactors are non-protein substances that enable proteins, including enzymes, to function. They can be either organic or inorganic. Prosthetic groups comprise a subset of cofactors that are permanently attached to the enzymes they assist. Other cofactors form temporary bonds with their enzymes–if organic, they are known as coenzymes. Most coenzymes are derived from vitamins. The following table summarises the characteristics of the different types of cofactor.

	Coenzyme	Inorganic cofactor	Prosthetic group
Are they organic (i.e. containing carbon) or inorganic?	Organic	Inorganic	Can be organic or inorganic
How do they interact with enzymes?	They form temporary bonds with the enzyme but leave following the reaction		Permanently bound to the enzyme
Examples	*Coenzyme A* (enzyme = acetyl CoA carboxylase) See 5.2.2(f) *NAD⁺* (enzyme = lactate dehydrogenase) See 5.2.2(f)	*Cl⁻* ions (enzyme = amylase) See 2.1.2(f) for the structure of amylose *Mg²⁺* (enzyme = DNA polymerase) See 2.1.3(e)	*Zn²⁺* (enzyme = carbonic anhydrase) See 3.1.2(i) *FAD* (enzyme = succinate dehydrogenase, an enzyme used in respiration) See 5.2.2

Precursor activation

Often enzymes are produced as inactive precursor proteins. The addition of a cofactor can activate an enzyme. The activation is usually achieved by altering the shape of the enzyme's active site.

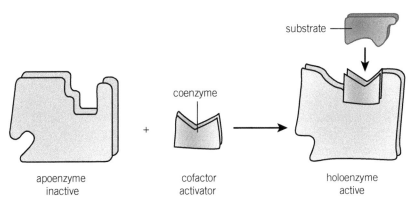

apoenzyme inactive   cofactor activator   holoenzyme active

▲ **Figure 1** *A cofactor can modify the shape of an active site*

Synoptic link

You learned about prosthetic groups in Topic 3.6, Structure of proteins.

Key term

Cofactor: A non-protein substance required for enzymes to function.

Summary questions

1 Outline the differences between coenzymes and prosthetic groups. *(2 marks)*

2 Suggest why some digestive enzymes are produced as inactive proteins. *(2 marks)*

3 Using your knowledge from earlier chapters, suggest how polypeptides are modified to produce enzymes with prosthetic groups. *(4 marks)*

Chapter 4 Practice questions

1   Which of the following is most likely to result in the denaturation of a protein?

   A The breaking of peptide bonds   **C** A small decrease in temperature

   B Competitive inhibition   **D** A small increase in pH   *(1 mark)*

2   A solution of amylase was diluted using the following method:

   1 ml of the stock solution was added to 9 ml of distilled water (forming a second solution). 2 ml of the second solution was then transferred to 8 ml of distilled water (forming a third solution). For the final dilution, 1 ml of the third solution was added to 19 ml of distilled water (forming a fourth solution).

   Which of the following represents the overall dilution from the stock solution of amylase to the fourth solution?

   A 3-fold dilution   **C** 1000-fold dilution

   B 100-fold dilution   **D** 5000-fold dilution   *(1 mark)*

3   Which of the following statements is true of non-competitive inhibitors?

   A They lower V_{max}

   B Their effects are always reversible

   C They bind to active sites

   D They alter the primary structure of an enzyme   *(1 mark)*

4   Which of the following statements is true of apoenzymes?

   A They can be activated by the addition of a prosthetic group

   B They are active precursors

   C The shapes of their active sites can be altered by coenzymes

   D They are formed from holoenzymes   *(1 mark)*

5   A group of students investigated the effect of pH on the activity of amylase. Their results are shown in the following table.

pH	Relative activity
4	30%
4.5	45%
6	85%
7	100%
9	65%

   a Based on the students' results, what is the optimum pH of amylase?

                                                          (1 mark)

   b State two improvements that the students could make to the presentation of their results.   *(2 marks)*

   c Suggest how the students could improve the validity of their results.

                                                           (1 mark)

6   The following passage describes the mechanism of competitive enzyme inhibition. Complete the passage by choosing the most appropriate word to place in each gap.

   A competitive inhibitor has a ……………. shape to the …............. ……......... of an enzyme and a ……………… shape to the enzyme's substrate.   *(3 marks)*

7   Suggest three factors that should be controlled when investigating the effect of substrate concentration on enzyme activity.   *(3 marks)*

5.1 The structure and function of membranes

Specification reference: 2.1.5(a) and (b)

The organelles and molecules discussed in previous chapters need to be organised and partitioned within cells. This organisation is the role of membranes, which we explore in this topic.

Roles of membranes

Membranes are located at cell surfaces (i.e. plasma membranes) and surrounding organelles (in eukaryotic cells). They have the following functions:

- **Physical barriers** (they separate intracellular environments from extracellular environments)
- They regulate the **exchange of substances** in and out of a cell; membranes are *partially permeable* (i.e. they allow only certain particles to pass through)
- **Compartmentalisation** (i.e. membranes enclose and isolate organelles, enabling them to maintain specific environments for chemical reactions)
- Support for the **cytoskeleton**
- Sites of **chemical reactions**
- Sites of **cell communication** (e.g. cell signalling)
- Formation of spheres called **vesicles**, which are used in bulk transport (see Topic 5.4, Active transport).

Membrane structure

Membranes consist primarily of a phospholipid bilayer, which has a hydrophobic core and hydrophilic phosphate heads. Various proteins are distributed within the bilayer. Membrane structure can be visualised as a fluid mosaic model ('fluid' because phospholipids are able to move within the bilayer, and 'mosaic' because proteins are scattered throughout the bilayer, like the tiles in a mosaic).

Synoptic link

You were introduced to phospholipids in Topic 3.5, Lipids.

▼ **Table 1** *Some of the components of membranes, in addition to phospholipids and cholesterol*

Component	Roles
Channel proteins	Facilitated diffusion (see Topic 5.3, Diffusion)
Carrier proteins	Facilitated diffusion and active transport (see Topics 5.3, Diffusion, and 5.4, Active transport)
Glycoproteins	Receptors (e.g. for neurotransmitters, peptide hormones, and drugs)
Glycolipids	Cell recognition (i.e. they act as antigens – see Chapter 12, Communicable diseases)

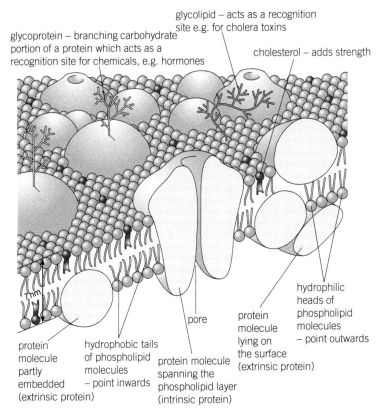

glycolipid – acts as a recognition site e.g. for cholera toxins

glycoprotein – branching carbohydrate portion of a protein which acts as a recognition site for chemicals, e.g. hormones

cholesterol – adds strength

hydrophilic heads of phospholipid molecules – point outwards

protein molecule lying on the surface (extrinsic protein)

protein molecule spanning the phospholipid layer (intrinsic protein)

pore

hydrophobic tails of phospholipid molecules – point inwards

protein molecule partly embedded (extrinsic protein)

7 nm

▲ **Figure 1** *The fluid mosaic model of membrane structure*

Go further: Cholesterol, a membrane buffer

Cholesterol is a lipid molecule that regulates the fluidity of membranes by inserting itself between phospholipids. The effect of the molecule, however, is temperature-dependent. At low temperatures cholesterol maintains fluidity by widening the gaps between phospholipids. Cholesterol prevents membranes becoming too fluid at high temperatures by attracting phospholipids and limiting their movement. Cholesterol therefore improves membrane stability.

1 Explain why cholesterol is sometimes referred to as a membrane 'buffer'.

2 Psychrophilic bacteria are adapted to cold habitats. Suggest the principal benefit of cholesterol in the plasma membranes of psychrophilic species.

Summary questions

1 Outline three roles of membranes *within* cells. (*3 marks*)

2 State and explain which type of protein within membranes would be used in the following cases: **a** The uptake of glucose into a cell **b** The binding of adrenaline to a liver cell. (*4 marks*)

3 Some cells, such as cells in the proximal convoluted tubules of the kidneys, require different proteins to be present on either side of the cell. Suggest why an increase in membrane fluidity would be a problem for these cells. (*2 marks*)

5.2 Factors affecting membrane structure

Specification reference: 2.1.5(c)

Membranes dictate the shape and content of cells and organelles. A loss of membrane structure would therefore damage the cell or organelle the membrane surrounds. Factors such as temperature and the presence of solvents determine the integrity of membrane structure.

Factors affecting membrane structure

Apart from shortages of membrane components, such as phospholipids and cholesterol, two factors can have a negative impact on membrane structure: temperature changes and the presence of solvents.

▼ **Table 1** *The effects of temperature and solvents on membrane structure*

Factor	Effect on membrane
Temperature	Decreased temperature reduces fluidity (phospholipids move less due to lower kinetic energy) Increased temperature causes greater fluidity and therefore increases permeability, but a membrane will lose its structure and break apart if temperature continues to rise
Solvents	Organic, weakly polar (e.g. ethanol) or non-polar (e.g. benzene) solvents disrupt or dissolve membranes, making them more permeable

water surface (outside of cell)

alcohol

water surface (inside of cell)

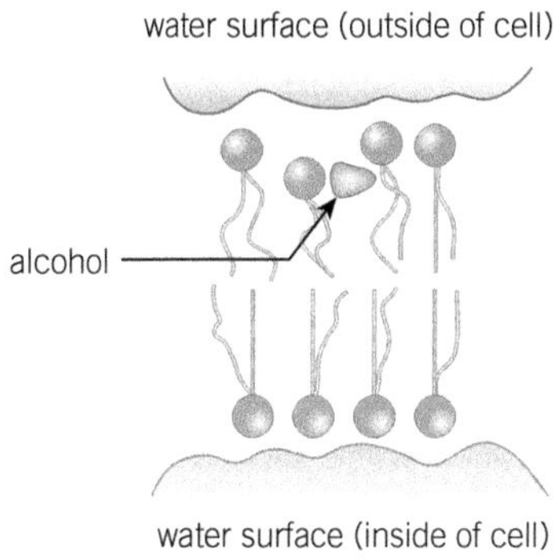

▲ **Figure 1** *Alcohol molecules (e.g. ethanol) disrupt membrane structure, thereby increasing permeability*

> **Revision tip: Membranes do not denature**
> Proteins within membranes can be denatured (see Topic 4.2, Factors affecting enzyme activity). Membranes as a whole (i.e. the phospholipid and cholesterol components) are not denatured by high temperatures. You can describe a membrane as being 'disrupted' or 'destroyed' by increases in temperature.

Practical skills: Investigating membrane permeability

Question and model answer

The effect of solvents, temperature, pH, and other factors on membrane permeability can be investigated experimentally.

The effect of alcohol, for example, can be studied using the following procedure:

Five test tubes of alcohol solution (0%, which is a control, 10%, 20%, 30%, and 40%) are set up. Beetroot cylinders are added to each tube. After 10 minutes, the colour of each alcohol solution is measured. A red pigment called betacyanin leaks from beet cells into the surrounding solution when membranes are damaged. The intensity of the colour released is proportional to the level of cellular damage.

1 State three factors the investigator should control in order to increase the validity of the experiment.
   Answers include: temperature, volume of alcohol solution, surface area: volume ratio of beet cylinders, pH.
2 The procedure is repeated three times. Explain how this improves the experiment.
   Answer: reproducibility is increased, and the spread of results can be assessed (using statistical tests).
3 Suggest how the colour of the alcohol solutions can be measured.
   Answer: using a colorimeter.

Summary questions

1 Describe how temperature affects the permeability of membranes. (*3 marks*)

2 Other than the effect on phospholipids, why else might membrane function be damaged at high temperatures? (*2 marks*)

3 Explain how excessive ethanol consumption might disrupt membrane function in cells. (*5 marks*)

5.3 Diffusion

Specification reference: 2.1.5(d)

Revision tip: Down, not across

Particles move *down* concentration gradients (from a high to low concentration) rather than *along* or *across* gradients.

Synoptic link

You will learn more about surface area to volume ratio and its influence on diffusion rate in Topic 7.1, Specialised exchange surfaces.

Summary questions

1   Glucose molecules are reabsorbed from proximal convoluted tubules in the kidney. Explain why the cell membranes of the proximal convoluted tubule: **a** are folded into microvilli; and **b** contain a high concentration of carrier proteins. (*4 marks*)

2   State how the following particles are likely to diffuse through a membrane. Explain your answers. **a** carbon dioxide **b** potassium ions. (*4 marks*)

3   Fick's law expresses diffusion rate in relation to surface area, membrane thickness, and concentration gradient. Complete the following equation, which represents a simplified version of Fick's law. Diffusion rate is proportional to _____ × _____ / _____. (*3 marks*)

One of the functions of membranes, as you learned in Topic 5.1, The structure and function of membranes, is to determine which substances move in and out of cells and organelles. The next three topics are concerned with the methods by which such transmembrane movement is achieved. You will begin by learning about passive transport.

Diffusion across membranes

Diffusion is the net (overall) movement of particles from a region of higher concentration to a region of lower concentration. It is a passive process (i.e. not requiring energy from ATP).

Diffusion through the bilayer

Some particles can diffuse through the bilayer, between phospholipid molecules. Large lipid-soluble molecules (e.g. steroid hormones), non-polar molecules (e.g. oxygen), and very small polar molecules (e.g. water) are able to pass directly through the bilayer.

Facilitated diffusion

Ions and large polar molecules (e.g. glucose and amino acids) pass through proteins rather than between phospholipids. The process of particles passing through transmembrane proteins is called facilitated diffusion. Two types of proteins are used:

Channel proteins: pores, which can be gated (i.e. opened and closed), allowing the diffusion of ions.

Carrier proteins: they have shapes that allow only the passage of specific molecules or ions.

Factors affecting diffusion rates

Higher temperatures increase diffusion rate by providing particles with more kinetic energy. In addition, the rate of diffusion increases with the following membrane characteristics:

- Steeper **concentration gradient** – the greater the difference in concentration either side of a membrane, the greater the diffusion rate
- Shorter **diffusion pathway** – thin membranes reduce the distance particles have to move
- Greater **surface area**
- A higher **concentration of carrier proteins** increases the rate of facilitated diffusion.

Practical skill: Investigating factors that affect diffusion rates

Dialysis (Visking) tubing is a partially permeable material that can be used to simulate a cell membrane. This tubing is often used in experiments to test the effect of concentration gradients or temperature on diffusion rate.

5.4 Active transport

Specification reference: 2.1.5(d) (i)–(ii)

The movement of particles down a concentration gradient does not require energy. You learned about this type of passive movement in Topic 5.3, Diffusion. Here we discuss movement against a concentration gradient, which requires energy. These particle movements across membranes are referred to as active transport.

Active transport

Energy is required for active transport because particles are being moved against a concentration gradient (i.e. from a region of lower to a region of higher concentration). The energy is in the form of ATP. Active transport can use carrier proteins as pumps or take the form of bulk transport.

Carrier proteins (pumps)

Carrier proteins change shape, thereby allowing particles to pass through them, due to ATP hydrolysis (which produces phosphate ions).

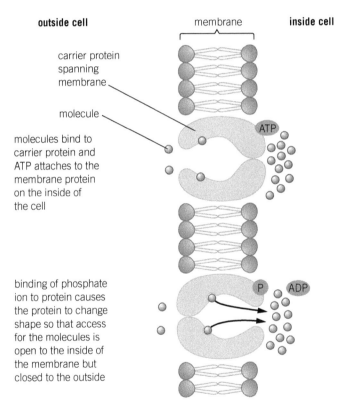

outside cell   membrane   **inside cell**

carrier protein spanning membrane

molecule

molecules bind to carrier protein and ATP attaches to the membrane protein on the inside of the cell

ATP

binding of phosphate ion to protein causes the protein to change shape so that access for the molecules is open to the inside of the membrane but closed to the outside

P   ADP

▲ **Figure 1** *The mechanism by which carrier proteins function*

Bulk transport

The movement of large molecules (e.g. enzymes and hormones) relies on bulk transport, which is the movement, in and out of cells, of particles within vesicles.

Endocytosis

Endocytosis is bulk transport *into* cells. Vesicles are formed by the plasma membrane being pinched off. Two forms exist:

Pinocytosis: a cell engulfs liquid and small dissolved particles

Phagocytosis: a cell engulfs large solid material (e.g. a white blood cell engulfing a bacterium).

Synoptic link

We discussed ATP as a molecule that transfers chemical energy in Topic 3.11, ATP.

Common misconception: Facilitated diffusion vs active transport

Carrier proteins are used for facilitated diffusion and active transport. In both processes, carrier proteins are selective (i.e. they change shape only when a specific particle binds to the carrier protein). Carrier proteins used for active transport, however, require ATP hydrolysis in addition to the binding of the particle. Facilitated diffusion does not require ATP.

Revision tip: Phagocytosis is not just for phagocytes

Phagocytosis is usually discussed in the context of phagocytes (cells of the immune system that take in and digest pathogens), but phagocytosis can occur in other cells.

Synoptic link

You will learn about the importance of phagocytosis as part of the immune system in Topic 12.5, Non-specific animal defences against pathogens.

Exocytosis

Exocytosis is bulk transport *out* of cells. In many cases, vesicles are formed by the Golgi apparatus and fuse with the cell surface membrane, releasing their contents (e.g. hormones).

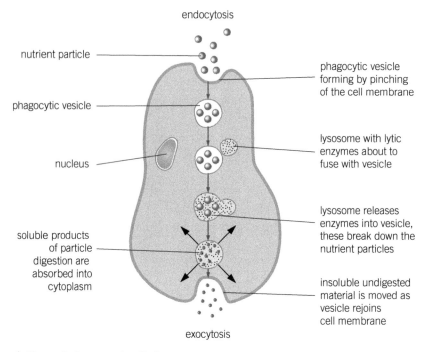

▲ **Figure 2** *An example of bulk transport*

Summary questions

1 Outline the characteristics of particles that are transported using exocytosis. *(4 marks)*

2 Complete the following table, which compares facilitated diffusion and active transport. *(5 marks)*

	Active transport	Facilitated diffusion
Uses carrier proteins		
Particles move down concentration gradients		
Particles move against concentration gradients		
Requires ATP		
At least two binding sites must be present on carrier proteins		

3 Suggest how the vesicles formed during phagocytosis would differ from those formed during pinocytosis. Explain your answer. *(3 marks)*

5.5 Osmosis

Specification reference: 2.1.5(e) (i)–(ii)

Osmosis and water potential

The direction in which water diffuses across a membrane is determined by water potential (ψ), which is measured in kilopascals (kPa). Water potential decreases as more solute particles dissolve. Osmosis always operates from a region of higher water potential to a region of lower water potential; in other words, water moves to the side of the membrane that has more dissolved solute.

Key term

Osmosis: Diffusion of water through a partially permeable membrane down a water potential gradient.

▼ **Table 1** *Comparison of the water potentials of pure water and glucose solutions*

	Water potential	Solute concentration
Pure water	Highest possible water potential (0 kPa)	No solute dissolved
Dilute glucose solution	High water potential (−20 kPa)	Low glucose concentration
Concentrated glucose solution	Low water potential (−400 kPa)	High glucose concentration

Revision tip: Nothing higher than zero

Pure water (without any solute dissolved) has a water potential of 0. This is the highest possible value of ψ. Solutions always have a negative water potential. As more solute dissolves, water potential drops.

Effects of osmosis on plant and animal cells

Osmosis produces different effects in plant and animal cells. These differences result from the cellulose walls surrounding plant cells.

▼ **Table 1** *The effects of osmosis on animal and plant cells*

Water potential of the solution surrounding the cell	Net movement of water	Effect on animal cell	Effect on plant cell
Higher (less solute in the surrounding solution)	Enters cell	Swells and bursts (i.e. the cell undergoes **lysis**)	Swells and becomes **turgid** (i.e. the cell is full of water and the membrane is pushed against the cell wall)
Equal (the same solute concentrations in the cell and surrounding solution)	Water leaves and enters the cell, but at equal rates	No change	No change
Lower (more solute in the surrounding solution)	Leaves cell	Shrinks (i.e. the cell undergoes **crenation**)	**Plasmolysis** (membrane pulls away from the cell wall)

haemoglobin is more concentrated, giving cell a darker appearance

cell shrunken and shrivelled

▲ **Figure 1** *Crenation (cell shrinkage) in an erythrocyte*

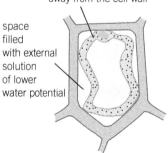

membrane completely pulled away from the cell wall

space filled with external solution of lower water potential

▲ **Figure 2** *Plasmolysis in a plant cell*

Summary questions

1. State whether the following sentences are true or false. **a** Diluting a solution will lower water potential. **b** No solutions can have a water potential of 0. **c** A liver cell is in danger of plasmolysing if placed in a solution with a high water potential. *(3 marks)*

2. An erythrocyte with a water potential of −500 kPa was placed in solutions with a range of water potentials. The water potentials are listed below. Describe what would happen to the erythrocyte in each case. **a** −470 kPa **b** −500 kPa **c** −1200 kPa *(3 marks)*

3. Suggest why some plants wilt when the soil they are growing in becomes too salty. *(4 marks)*

1 Which of the following sentences is the most accurate description of the role of cholesterol within membranes?

   A Cholesterol increases membrane fluidity

   B Cholesterol limits the fluidity of a membrane at low temperatures

   C Cholesterol regulates membrane fluidity

   D Cholesterol prevents membrane fluidity from increasing   *(1 mark)*

2 Which of the following substances is an example of a non-polar solvent?

   A Benzene

   B Water

   C Cholesterol

   D Phospholipid   *(1 mark)*

3 Which of the following statements is true of facilitated diffusion?

   A Particles move up a concentration gradient (from a low to a high concentration)

   B Only occurs through channel proteins

   C Does not require ATP

   D Rate is independent of temperature   *(1 mark)*

4 Which is the most appropriate term to describe the release of neurotransmitters from a neuronal cell?

   A Endocytosis

   B Exocytosis

   C Pinocytosis

   D Phagocytosis   *(1 mark)*

5 Explain how the structure of membranes enables them to perform the following functions:

   • Cell communication

   • The regulation of the exchange of substances in and out of organelles.   *(5 marks)*

6 Explain why membranes are described using a fluid mosaic model.   *(2 marks)*

7 The following passage describes the movement of water via osmosis. Complete the passage by choosing the most appropriate word to place in each gap.

Water diffuses across a membrane from the side with the .... water potential to the side with the .... water potential. Plant cells that lose water will exhibit reduced .... and may undergo .... (i.e. the plasma membrane pulls away from the cell wall).

   (4 marks)

8 Describe and explain the movement of water between the following three cells, which share boundaries with each other.

Cell A (water potential of −250 kPa)

Cell B (water potential of −180 kPa)

Cell C (water potential of −280 kPa)   *(3 marks)*

6.1 The cell cycle

Specification reference: 2.1.6(a) and (b)

Having learned about the structure and components of cells in previous chapters, you will now explore how cells divide and differentiate. We begin in this topic by examining how a sequence of events, known as the cell cycle, enables a cell to divide.

The phases of the cell cycle

Before dividing, cells must grow and synthesise new organelles and molecules. These processes occur in a particular order, which is known as the cell cycle. Cells are in interphase, which is when growth and synthesis occurs, for the majority of the cycle. Mitosis (division of the cell nucleus) and cytokinesis (cell division) (see Topic 6.2, Mitosis) follow interphase. The key features of the cell cycle phases are illustrated in the following diagram.

Synoptic link

You learned about DNA replication in Topic 3.9, DNA replication and the genetic code.

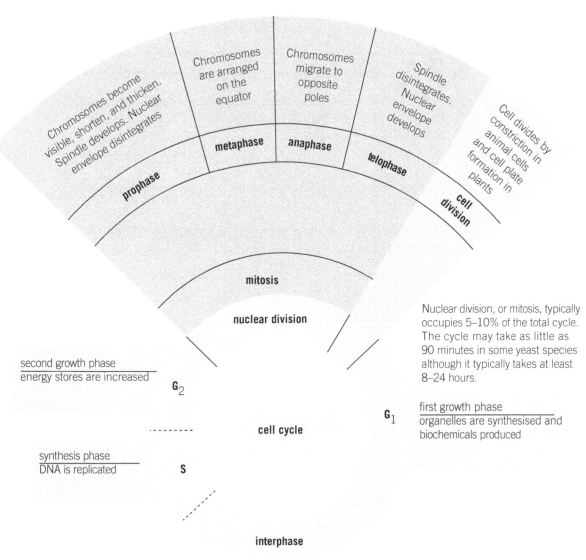

Nuclear division, or mitosis, typically occupies 5–10% of the total cycle. The cycle may take as little as 90 minutes in some yeast species although it typically takes at least 8–24 hours.

▲ **Figure 1** *Interphase comprises the majority of the cell cycle. The duration of one cycle varies between species and cell types*

Interphase is sometimes referred to as the 'resting phase'. This is misleading; in fact, it could not be further from the truth. A great deal of chemical activity, including the synthesis of organelles and DNA, takes place during interphase.

However, cells sometimes leave the cell cycle and stop dividing. For example, some mature, differentiated cells (see Topic 6.4, The organisation and specialisation of cells), such as neurons, no longer divide. The cycle can also be halted in cells with damaged DNA. These cells are said to have entered the G_0 phase. 'Resting phase' would be a more appropriate label for the G_0 phase rather than interphase.

Cell cycle control

The order and timing of processes in the cell cycle are under tight control. The cycle has checkpoints (e.g. at the end of the G_1 and G_2 phases) that verify whether each phase of the cycle has been completed correctly. These checkpoints are controlled by proteins called cyclins and cyclin-dependent kinases. The cell cycle can be halted when errors are detected at a checkpoint.

Go further: How can we analyse the cell cycle?

A technique called flow cytometry can be used to determine the length of each phase in an organism's cell cycle. This will differ between species. A calculation that can be used for estimating the length of a phase is:

$$length\ of\ phase = \frac{T_c \times \ln(fp + 1)}{\ln 2}$$

[Where T_c = cell cycle duration, fp = the fraction of cells in a phase, ln = natural logarithm]

For example, imagine a cell cycle that lasts 24 hours, in which 30% of cells are found to be in the S phase. You can calculate the length of the S phase as follows:

$$length\ of\ phase = \frac{24 \times \ln(0.3 + 1)}{\ln 2} = \frac{24 \times 0.26}{0.69} = 9.0\ hours$$

1   Calculate the length of the G_1 phase in a cell cycle that lasts 18 hours and in which 45% of cells are in the G_1 phase.

2   Changes in the length of cell cycle phases can indicate health problems. Suggest what problems these changes in cycle length might indicate.

Summary questions

1   State what is produced during **a** the G_1 phase of interphase **b** the S phase of interphase.   (*3 marks*)

2   Explain the importance of checkpoints during the cell cycle.   (*3 marks*)

3 a   Suggest why it would be inappropriate to use the term 'resting phase' in relation to a cell in the cell cycle.

   b   Suggest which types of cells could be considered to be in a 'resting phase'.   (*3 marks*)

6.2 Mitosis

Specification reference: 2.1.6 (c), (d), and (e)

You learnt in Topic 6.1, The cell cycle, that cells spend most of their time growing and synthesising molecules, in addition to performing specialised functions. The division of a cell takes up a relatively small proportion of its cycle, yet it is a crucial process. Here you will examine the steps involved in nuclear division (mitosis) and cell division (cytokinesis).

The importance of mitosis

Mitosis produces two nuclei that contain identical genetic material. This ensures that, following cell division, the two daughter cells are exact copies of the original cell. The replication of genetically identical cells is important for the following processes:

- *Growth*
- *Repair* (of damaged cells)
- *Replacement* (of cells, such as red blood cells, that have limited lifespans)
- *Asexual reproduction* (in eukaryotes).

The stages of mitosis

Mitosis is preceded by interphase. DNA replicates during interphase; therefore a cell at the start of mitosis has two copies of all its genetic material (i.e. two identical DNA molecules, called chromatids).

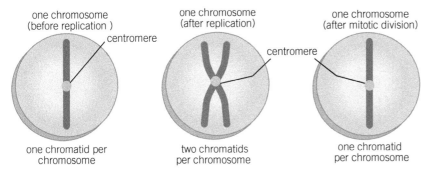

▲ **Figure 1** *Chromosome structure at different stages of the cell cycle*

Cell division

Revision tip: Mitosis in plants

Mitosis is different in plants and animals.

Mitosis in animals	Mitosis in plants
Centrioles	No centrioles (in most plants)
Cells become rounded before division	No shape change
Spindle disappears before cytokinesis	Some of the spindle remains during cytokinesis

Summary questions

1 Human somatic cells have a diploid number of 46. State the number of chromatids present in a human cell **a** during prophase **b** at the end of cytokinesis.   (*2 marks*)

2 Plants lack centrioles. What does this indicate about the role of centrioles in mitosis?
(*2 marks*)

3 The calculation of mitotic index provides an indication of the rate of cell division in a tissue. *Mitotic index =*
number of cells in mitosis
total number of cells
Suggest which disease might result in an increase in mitotic index. Explain your answer.   (*2 marks*)

Stage	Key features	Appearance
Prophase	• **Chromatin** (comprising DNA and histone proteins) condenses – **chromosomes become visible.** • **Nucleolus** disappears. • **Centrioles** move to the poles of the cell. • **Nuclear envelope** breaks down (towards the end of prophase).	nuclear envelope disintegrates; spindle microtubules
Metaphase	• **Spindle fibres** (organised by the centrioles) attach to **centromeres** (towards the end of prophase/beginning of metaphase). • **Chromosomes line up** along the centre (equator) of the cell.	metaphase plate equator; mitotic spindle
Anaphase	• Spindle fibres shorten. • **Centromeres divide.** • **Chromatids are separated** and pulled to opposite poles of the cell.	sister chromatids separate
Telophase	• Chromatids are at the poles of the cell (and can be referred to as daughter chromosomes). • **Nuclear envelopes reform** around each set of chromosomes. • **Chromosomes uncoil** (and are no longer visible). • Cell division (**cytokinesis**) begins.	

Cytokinesis

The division of a cell (cytokinesis) begins in telophase. This results in two genetically identical cells. Each cell receives approximately half of the organelles and cytoplasm from the original cell.

Different methods of cytokinesis operate in animal and plant cells. This is principally a consequence of plants having cell walls.

In **animals:** a *cleavage furrow* forms (i.e. cell surface membranes are pulled together by the *cytoskeleton*).

In **plants:** the cell wall prevents cleavage furrows. The two daughter cells are instead separated by new *cell wall production* down the centre of the original cell.

6.3 Meiosis

Specification reference: 2.1.6(f) and (g)

Mitosis, as you learned in Topic 6.2, Mitosis, is nuclear division that produces two identical nuclei. The formation of reproductive cells, however, creates genetic variation and involves a different type of nuclear division called meiosis.

The stages of meiosis

Meiosis I separates homologous chromosomes to produce two haploid cells. Meiosis II is similar to mitosis: chromatids are separated to form a total of four haploid daughter cells.

Stage	Key features	Appearance
Prophase I	The key events of mitotic prophase (see Topic 6.2, Mitosis) occur (i.e. nuclear envelope disintegrates, nucleolus disappears, spindles form, chromosomes condense). Homologous chromosomes pair up (form **bivalents**). **Crossing over**	nuclear envelope — spindle microtubules and centriole
Metaphase I	Homologous pairs (bivalents) line up at the cell equator. **Independent assortment** of chromosomes.	bivalents aligned on the equator
Anaphase I	Homologous chromosome pairs are separated (**random segregation**) – sister chromatids both remain attached to centromeres.	homologous chromosomes being pulled to opposite poles
Telophase I	Chromosomes assemble at either pole. **Cytokinesis** (cell division) – two haploid cells formed.	
Prophase II	Chromosomes condense, nuclear envelope breaks down, and spindles form.	
Metaphase II	Chromosomes line up at equator (as in mitosis). **Independent assortment** of chromatids.	

Revision tip: Homologous chromosomes

Each nucleus contains pairs of matching chromosomes, which have the same gene loci (but possibly different versions of those genes, i.e. different alleles). These pairs are known as homologous chromosomes. Each pair comprises a maternal chromosome and a paternal chromosome. Homologous pairs are the same length and, at the start of nuclear division, each chromosome consists of two chromatids.

Key term

Meiosis: A form of nuclear division in which the chromosome number is halved.

Synoptic link

We explored DNA structure in Topic 3.9, DNA replication and the genetic code.

Anaphase II	**Random segregation** of chromatids.	
Telophase II	Chromatids assemble at poles. **Cytokinesis** (cell division) – four haploid daughter cells formed.	

Summary questions

1 Describe the differences between **a** mitosis and meiosis II **b** meiosis I and meiosis II. *(4 marks)*

2 The number of possible combinations of chromosomes in a gamete can be calculated using the term 2^n, with 'n' being the haploid number of the species. Tigers (*Panthera tigris*) have a diploid chromosome number of 38. Calculate the number of different chromosome combinations that are possible in a tiger gamete. *(1 mark)*

3 Explain how genetic variation is introduced to gametes during meiosis. *(4 marks)*

The importance of meiosis

Species that undergo asexual reproduction produce genetically identical offspring. Many multicellular organisms, however, reproduce sexually. These species produce sex cells (gametes) that contain half the standard number of chromosomes (i.e. the haploid number). Meiosis, which is known as a reduction division, enables the chromosome number to be halved. As a result, the standard chromosome number (i.e. the diploid number) is restored in the offspring when two organisms reproduce.

As well as ensuring the preservation of chromosome number through generations, meiosis introduces genetic variation. This variation is introduced through the processes outlined in the following table:

Process	Stage of meiosis	What happens?	How does this produce variation?
Crossing over	Prophase I	Non-sister chromatids (chromatids from different chromosomes in a homologous pair) interweave (at points called chiasmata), forming bivalents.	Genetic material is exchanged between homologous chromosomes, producing new combinations of alleles.
Independent assortment of chromosomes	Metaphase I	When homologous chromosomes move to the cell equator, the alignment of each chromosome (i.e. on which side of the cell it is positioned) is random.	When homologous chromosomes are separated in anaphase I, many different chromosome combinations can be formed in daughter cells.
Independent assortment of chromatids	Metaphase II	Chromosomes line up at the cell equator. The side on which each sister chromatid is positioned is random.	Sister chromatids are no longer identical (due to crossing over), therefore many chromatid combinations are possible in daughter cells.

6.4 The organisation and specialisation of cells

Specification reference: 2.1.6(h) and (i)

Although they share common features, cells in an organism are not identical; different cells become specialised for particular roles, depending on their positions within the organism.

Specialised cells

In order to specialise for a particular function, cells differentiate. All cells in an organism contain the same DNA; differentiation requires a cell to express certain genes but switch off others. The table summarises the adaptations exhibited by some specialised cells.

	Structural feature	Function
Red blood cells/ erythrocytes	Flattened, biconcave shape.	Increases the surface area: volume ratio to increase the rate of O_2 diffusion.
	No organelles.	More space available for haemoglobin.
	Flexible due to protein arrangements in membranes.	Ability to squeeze through capillaries.
Neutrophils (a type of white blood cell)	Multi-lobed nucleus.	Ability to squeeze through gaps in capillary walls to reach infections.
	Many lysosomes.	They contain hydrolytic enzymes (to destroy pathogens).
Sperm cells	Flagellum (tail).	For movement towards the egg.
	Many mitochondria.	To supply energy for movement.
	Acrosome.	Contains digestive enzymes to enable the sperm to penetrate the egg.
	haploid nucleus / acrosome / mid-piece (7 μm long) / tail (40 μm long, two-thirds of it omitted from this drawing) / centriole / microtubules (in a 9+2 arrangement) / protein fibres to strengthen the tail / plasma membrane / helical mitochondria	
Palisade cells	Many chloroplasts (packed close together).	High rate of light absorption for photosynthesis.
	Thin cell walls.	Greater CO_2 diffusion rate.
	Large vacuole.	To maintain turgor pressure.
Root hair cells	Long, narrow extensions of the cell.	Large surface area to increase water and mineral uptake from the soil.
	Large vacuole with a high concentration of dissolved solutes.	Lower water potential to increase rate of water uptake from the soil.
Guard cells	Two kidney-shaped cells with thickened inner cell walls	They control when stomata open (depending on the requirement for gas exchange and water levels in the plant)
	chloroplast / cytoplasm / vacuole / nucleus / thin outer wall / thick inner wall	

Synoptic link

You will learn more about xylem and phloem tissue in Chapter 9, Transport in plants.

Tissues

A tissue is a collection of differentiated cells that work together for a particular function. Examples include:

Tissue	Location	Structural features	Function
Squamous epithelium	Capillaries and alveoli	Thin/flat	Increases diffusion rate
Ciliated epithelium	Trachea	Cilia on the outside of cells	Sweep mucus from trachea
Cartilage	Joints (between bones)	Firm and flexible	Protective connective tissue
Skeletal muscle	Attached to bones	Contractile proteins	Movement of the skeleton
Xylem	Plant stem	Elongated dead cells strengthened by lignin	Transport of water
Phloem	Plant stem	Perforated walls	Transport of nutrients

Organs

Organs are collections of several tissues that combine to perform a function or range of functions. For example, the heart comprises squamous epithelium, endothelium, and cardiac muscle, as well as other tissues.

Organ systems

Organs that work in conjunction with each other can be considered an organ system. For example, the digestive system includes the following organs: oesophagus, stomach, liver, pancreas, gall bladder, gastrointestinal tract.

Overall, an organisational hierarchy exists:

Specialised cells – tissues – organs – organ systems – whole organism.

Summary questions

1 Outline how a sperm cell is specialised for its function. (*4 marks*)

2 Explain how the presence of squamous epithelium tissue improves the efficiency of gas exchange in the lungs. (*4 marks*)

3 Sclerenchyma tissue provides support to plants. Suggest some features of sclerenchyma tissue that enable it to carry out its function. (*3 marks*)

6.5 Stem cells

You looked at examples of differentiated cells in Topic 6.4, The organisation and specialisation of cells. Here we discuss the undifferentiated cells from which specialised cells arise. These undifferentiated cells are known as stem cells.

What are stem cells?

Although they are undifferentiated, stem cells vary in the range of cells they have the potential to form; this is referred to as their potency. The following table outlines the traits of different types of stem cells.

Potency of stem cell	Which cells can they divide to form?	Examples
Totipotent	Any (and they have the potential to form a whole organism)	The first 16 cells of an animal zygote. Plant meristem cells (including cambium tissue, which differentiates into xylem and phloem tissue).
Pluripotent	All tissues (but not a whole organism)	Early embryonic cells (in the blastocyst)
Multipotent	A limited range of cells	Haematopoietic stem cells (in bone marrow), which can differentiate to form all blood cells, including erythrocytes and neutrophils

Uses of stem cells

Current uses include:

- Drug testing *in vitro* (i.e. testing drugs on cultured cells in a laboratory)
- Studying developmental biology and disease development *in vitro*
- Treatment of burns
- Bone marrow transplants to replace stem cells destroyed during cancer treatment.

Future diseases that could be treated include: Parkinson's and Alzheimer's disease (using neural stem cells), heart disease, and type 1 diabetes.

Ethics

Some people have moral and religious reservations about the use of pluripotent stem cells obtained from embryos. Scientists are developing techniques for reprogramming differentiated adult cells back into pluripotent stem cells (known as induced pluripotent stem cells (iPSCs)). These techniques do not require the use of embryos and would therefore overcome the ethical objections held by some people.

Summary questions

1 Describe one similarity and one difference between totipotent and pluripotent stem cells.
(*2 marks*)

2 Explain the potential benefits of using induced pluripotent stem cells rather than pluripotent embryonic stem cells for disease treatment in the future. (*3 marks*)

3 Suggest why plants are more able than animals to form reproductive clones. (*3 marks*)

1 Crossing over during meiosis contributes to genetic variation.

Which is the correct description of crossing over?

A Chiasmata form when chromatids on non-homologous chromosomes exchange genetic material in prophase I.

B Chiasmata form when chromatids on homologous chromosomes exchange genetic material in prophase I.

C Chiasmata form when chromatids on non-homologous chromosomes exchange genetic material in metaphase I.

D Chiasmata form when chromatids on homologous chromosomes exchange genetic material in metaphase I. *(1 mark)*

2 Mesenchymal stem cells can differentiate into adipose, bone, cartilage, and connective tissue cells, but not other types of cell.

What type of stem cells are mesenchymal stem cells?

A Unipotent   C Pluripotent

B Multipotent   D Totipotent   *(1 mark)*

3 One method for estimating the length of a cell cycle phase is:

$$\text{Length of phase} = \frac{[T_C \times \ln(fp + 1)]}{\ln 2}$$

(where T_C = cell cycle duration, fp = the fraction of cells in a phase, ln = natural logarithm)

What is the length of the G_2 phase in a cell cycle that lasts 16 hours and in which 35% of cells are in the G_2 phase?

A 3 hours   C 7 hours

B 6 hours   D 8 hours   *(1 mark)*

4 Describe the potential uses of stem cells in disease treatment and the problems that have been encountered when researching potential stem cell therapies. *(6 marks)*

5 The following passage describes prophase I of mitosis. Complete the passage by choosing the most appropriate word to place in each gap.

.... (i.e. DNA and its associated proteins) condenses in prophase I. The .... and nuclear membrane disappear. .... move to opposite poles of the cell. These structures are composed of .... *(4 marks)*

6 The letters A–D list features of specialised cells. Match each of the features to the specialised cell (W–Z) being described. In each case, explain the benefit of the adaptation.

A Thin cell walls   W Neutrophil

B Acrosome   X Erythrocyte

C No organelles   Y Sperm cell

D Many lysosomes   Z Palisade cell   *(8 marks)*

7.1 Specialised exchange surfaces

Specification reference: 3.1.1(a) and (b)

Single-celled organisms obtain oxygen through diffusion, which you learnt about in Topic 5.3, Diffusion. Larger, multicellular organisms, however, cannot rely on diffusion alone; they have evolved specialised gas exchange surfaces.

The need for specialised exchange surfaces

Multicellular organisms have evolved exchange surfaces because:

- Metabolic activity is greater in multicellular organisms, which means oxygen needs to be supplied and carbon dioxide removed at higher rates.
- Surface area to volume (SA:V) ratios are smaller in multicellular organisms (i.e. volume increases in relation to surface area), which means that diffusion alone would not achieve an adequate rate of gas exchange.

Features of exchange surfaces

We will explore examples of exchange surfaces that have evolved in different species in Topic 7.2, The mammalian gaseous exchange system, and Topic 7.4, Ventilation and gas exchange in other organisms. All of these adaptations have features in common.

▼ **Table 1** *Common features of gas exchange surfaces*

Feature of exchange surfaces	Benefit
Increased surface area	Overcomes the reduced SA:V ratio in larger organisms
Thin layers	Reduces the diffusion distance
Good blood supply	Maintains steep concentration gradients through the quick supply and removal of gases
Good ventilation	

Synoptic link

You learned about the passive movement of substances in Topic 5.3, Diffusion.

Maths skill: SA:V ratios

Surface area to volume ratio decreases as organisms increase in size. We can demonstrate this effect by tracking the changes in surface area and volume of spheres as they increase in size.

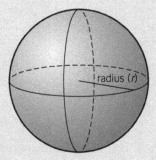

radius (*r*)

Surface area of a sphere = $4\pi r^2$

Volume of a sphere = $\frac{4}{3}\pi r^3$

For example, a sphere with a radius of 1 m would have a surface area of 12.6 m² (4 × 3.14 × 1 × 1) and a volume of 4.2 m³ (4/3 × 3.14 × 1 × 1 × 1). When the radius is increased to 2 m, the proportional difference between the surface area (50.2 m²) and volume (33.5 m³) decreases. The SA:V ratio becomes smaller.

Summary questions

1. Explain why multicellular organisms require specialised gas exchange surfaces. *(4 marks)*

2. Describe how specialised gas exchange surfaces maintain concentration gradients to enable a high rate of diffusion. *(4 marks)*

3. Actinophryids are spherical microorganisms. A population of one genus of actinophryids (*Actinophrys*) has a mean diameter of 45 μm. Another genus (*Actinosphaerium*) has a mean diameter of 400 μm. Calculate the SA:V ratios for the two genera. *(6 marks)*

7.2 The mammalian gaseous exchange system
7.3 Measuring the process

Specification reference: 3.1.1(c), (d), and (e)

Mammals have evolved a gaseous exchange system that performs a balancing act: limiting the amount of water lost from the body while enabling efficient gas exchange. The exchange of gases in mammals occurs in the lungs, the features of which are our focus in this topic.

Structures in the mammalian gas exchange system

The mammalian gas exchange system consists of passages (the nasal cavity, trachea, bronchi, and bronchioles) and the gas exchange surfaces (alveoli in the lungs) to which they deliver air. The principal features of these structures are outlined in the following table.

Structure	Key features	Function
Nasal cavity	Good blood supply	Warms air entering the body
	Lined with hairs and **mucus**-secreting cells	Traps dust and bacteria (protection from disease)
	Moist surface	Increases humidity, reducing evaporation from the lungs
Trachea	Supported by flexible **cartilage**	Prevents collapse
	Lined with **goblet cells**, which secrete mucus	Trap dust and bacteria
	Ciliated epithelium cells	Cilia move mucus away from the lungs
Bronchus	Cartilage, like the trachea	Prevents collapse
Bronchioles	**Smooth muscle** (and no cartilage)	Bronchioles can constrict and dilate to vary the amount of air reaching the lungs
	Flattened epithelium cells	Some gaseous exchange is possible
Alveoli	Single layer of **flattened epithelium** cells	Short diffusion pathway, which increases diffusion rate
	Elastic fibres and **collagen**	Enable stretching and elastic recoil during ventilation
	Large surface area	Increased rate of diffusion (*see Topic 5.3, Diffusion, and Topic 7.1, Specialised exchange surfaces*)
	Good blood supply (alveoli are surrounded by a capillary network) and good ventilation	O_2 is supplied to the alveoli and moved into the circulatory system quickly. CO_2 is supplied from the circulatory system and removed from the lungs quickly. This maintains a **steep concentration gradient**.
	Covered with a layer of **surfactant**	Alveoli remain inflated

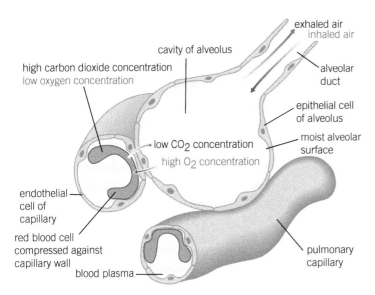

▲ **Figure 1** *Alveolar gas exchange*

Mechanism of ventilation

Air moves in and out of the lungs because of changes in pressure; this is known as ventilation. Efficient ventilation (i.e. the efficient supply of oxygen to the alveoli and the quick removal of carbon dioxide) maintains a steep concentration gradient in the lungs. The following table compares the changes that occur in the lungs during inspiration and expiration.

Inspiration (inhalation)	Expiration (exhalation)
External intercostal muscles contract	External intercostal muscles relax
Ribs move up and out	Ribs move down and inwards
Diaphragm contracts and flattens	Diaphragm relaxes and reverts to its domed shape
Thorax volume increases	Thorax volume decreases
Air pressure in the lungs drops below atmospheric pressure	Air pressure in the lungs rises above atmospheric pressure
Air moves into the lungs	Air moves out of the lungs

Common misconception: Relax, you're exhaling normally …

Inspiration is an active process (i.e. external intercostal muscles and the diaphragm contract). Normal expiration (breathing out when at rest) is passive; the diaphragm and external intercostal muscles relax and return to their resting positions. Forceful exhalation (e.g. during physical exertion or coughing), however, requires muscular contraction to move air from the lungs; the internal intercostal muscles contract, pulling the ribs down, and abdominal muscles force the diaphragm back to its domed position.

How is lung function measured?

Measurements of lung function tend to be made using two pieces of equipment: a peak flow meter and a spirometer.

Equipment	Method	Use
Peak flow meter	Measures the rate at which patients expel air into a handheld tube.	Can be used to monitor conditions such as asthma.
Spirometer	Patients breathe in and out of a mouthpiece attached to a sealed chamber; oxygen from the chamber is used up.	Can measure several aspects of lung volume (see below).

Worked example: Breathing calculations

Breathing rate = the number of breaths taken per minute.

Ventilation rate (the total volume inhaled per minute) = breathing rate × tidal volume

You may be asked to rearrange this equation. For example, what is the tidal volume of a person with a ventilation rate of 6.5 dm³ min⁻¹ and a breathing rate of 13 breaths min⁻¹?

Tidal volume = ventilation rate / breathing rate = 6.5 / 13 = 0.5 dm³ (500 cm³).

What is measured in spirometry?

A variety of measurements can be made using a spirometer.

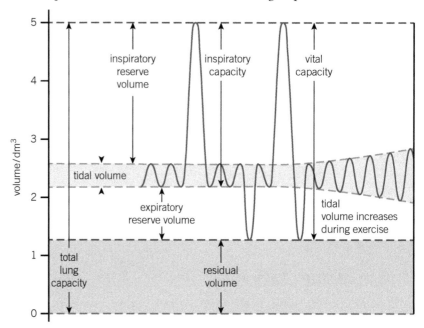

▲ **Figure 2** *Lung capacity has many measurable components*

- **Total lung capacity** = vital capacity + residual volume.
- **Residual volume** is the volume remaining in the lungs even after a person has exhaled with maximum force.
- **Vital capacity** is the maximum volume that can be breathed out following the strongest possible inhalation (i.e. tidal volume + inspiratory reserve volume + expiratory reserve volume).
- **Tidal volume** is the volume inhaled with each resting breath (or the volume exhaled with each resting breath).
- Inspiratory reserve and expiratory **reserve volumes** are the additional volumes of air that can be breathed in and out during forced inhalation and exhalation.

Summary questions

1 State three features of alveoli that enable a high rate of gas exchange. *(3 marks)*

2 Calculate the breathing rate of a person with a ventilation rate of 6.20 dm³ min⁻¹ and a tidal volume of 0.52 dm³. Give your answer to two significant figures. *(2 marks)*

3 The partial pressure (kPa) of a gas reflects its relative concentration in a gas mixture. The table below shows partial pressures of oxygen and carbon dioxide in three parts of the mammalian body.

Gas	Alveolar air (kPa)	Blood in pulmonary artery (kPa)	Blood in pulmonary vein (kPa)
Oxygen	13.8	5.3	13.8
Carbon dioxide	5.3	6.1	5.3

   a Explain why blood in the pulmonary vein has the same partial pressures as the air in the alveoli. *(2 marks)*
   b Suggest an explanation for the differences in partial pressure of oxygen and carbon dioxide in the pulmonary artery and alveolar air. *(2 marks)*

7.4 Ventilation and gas exchange in other organisms

Specification reference: 3.1.1(f), (g), and (h)

In Topic 7.2, The mammalian gaseous exchange system, you read about the lungs, which are the organs of gas exchange in mammals. Different internal gas exchange systems have evolved in other groups of multicellular organisms. Here we discuss two of these systems: the tracheal system in insects and gills in bony fish.

Gaseous exchange in insects and fish

Insects evolved hard exoskeletons for structural support and to limit water loss. These exoskeletons prevent the diffusion of gas across insects' body surfaces. Insects also lack blood pigments capable of transporting oxygen. As a consequence, insects evolved a system of tubes (**tracheae**) that deliver oxygen directly to cells (and remove carbon dioxide).

Aquatic organisms have another difficulty to overcome. Water is considerably denser than air; moving water in and out of a lung-like structure would require too much energy. Instead, bony fish have evolved structures called **gills**, which extract oxygen from water flowing across them in one direction.

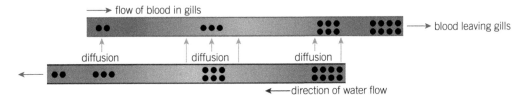

▲ **Figure 1** *Countercurrent flow in the gills. An oxygen concentration gradient is maintained. (Circles represent arbitrary units of oxygen concentration)*

> ### Revision tip: Examining gas exchange systems
> You may have the opportunity to dissect and examine the structures discussed here. You should be able to interpret diagrams and photomicrographs of the gill and tracheal tissue. Revising microscopy skills (Topic 2.1, Microscopy, and Topic 2.3, More microscopy) and magnification calculations (Topic 2.2, Magnification and calibration) will help.

> ### Revision tip: Tracheae
> You will have encountered tracheae in the context of mammalian gas exchange. *Trachea* means 'air pipe'; it is a tube that carries gases in and out of organisms. Mammals have one (singular: trachea), whereas insects have many (plural: tracheae).

▼ **Table 1** *A comparison of insect and fish gas exchange systems*

	Insect tracheae	Gills in bony fish
How does air/ water enter and leave the organism?	Air enters and leaves through small openings called **spiracles**. Spiracles are closed when possible to minimise water loss.	Water enters through the mouth. A continuous flow of water across the gills is achieved by the floor of the buccal cavity (mouth) being lowered (which takes water into the mouth) then raised (which forces water over the gills). Water leaves a fish when the operculum (a flap covering the gills) opens.
How is oxygen taken up?	Air travels through tracheae, which branch into tracheoles. Oxygen diffuses into cells from the tracheoles. Some active insects have high energy demands; they have evolved a muscular pumping system to increase oxygen supply.	Oxygen diffuses into gill **lamellae**, which are thin plates, packed with blood capillaries. Lamellae are attached to gill **filaments**. Water and blood (in lamellae) flow in opposite directions (i.e. **countercurrent flow**). This ensures an oxygen concentration gradient is maintained along the gill.
How does carbon dioxide leave?	CO_2 diffuses from tissues into the tracheae down a concentration gradient.	CO_2 diffuses from capillaries, across lamellae, and into water, which leaves through the operculum.
Appearance of the gas exchange system		

Summary questions

1 Explain why the evolution of a specialised gas exchange system was necessary for insects. *(2 marks)*

2 Compare the adaptations of alveoli and gills for gas exchange. *(6 marks)*

3 Explain how the structure of gills enables concentration gradients to be maintained during gas exchange. *(3 marks)*

Chapter 7 Practice questions

1 Which of the following is a prominent feature of the bronchioles?

    A Smooth muscle

    B Cartilage

    C Goblet cells

    D Ciliated epithelium   *(1 mark)*

2 Which of the following is the correct description of tidal volume?

    A The maximum volume of air that can be inhaled in addition to resting inhalation.

    B The maximum volume of air that can be exhaled in addition to resting exhalation.

    C The volume of air inhaled with each resting breath.

    D The volume of air that remains in the lungs after forced exhalation.

          (1 mark)

3 The breathing rate of a person is measured as 15 breaths per minute. The person's ventilation rate is $6.4\,dm^3$ per minute. Which of the following values represents the person's tidal volume?

    A $2\,cm^3$

    B $96\,cm^3$

    C $427\,cm^3$

    D $96\,000\,cm^3$ *(1 mark)*

4 Air passes through several structures during gas exchange in insects. Which of the following sets of words represent the order in which air passes through the insect exchange system?

    A Spiracle, tracheae, tracheoles

    B Spiracle, tracheoles, tracheae

    C Tracheae, tracheoles, spiracle

    D Tracheoles, tracheae, spiracle   *(1 mark)*

5 **a** A cell of the bacterial species *Streptococcus pyogenes* was analysed. The diameter of the cell was found to be $1\,\mu m$. Calculate the surface area to volume ratio of the cell.   *(4 marks)*

    b Explain why *Streptococcus pyogenes* does not require a specialised gas exchange system.   *(2 marks)*

6 The following passage describes the inhalation of air in mammalian lungs. Complete the passage by choosing the most appropriate word to place in each gap.

    The .... intercostal muscles contract; this causes the ribs to move up and out. The diaphragm contracts and .... The volume of the .... increases, and air .... in the lungs decreases below that of the atmosphere.   *(4 marks)*

8.1 Transport systems in multicellular animals

Specification reference: 3.1.2(a) and (b)

Synoptic link

You considered the concept of surface area to volume ratio in Topic 7.1, Specialised exchange surfaces.

Revision tip: Gas transport in insects

The tracheal system (see Topic 7.4, Ventilation and gas exchange in other organisms) is the principal site of oxygen and carbon dioxide exchange in insects. However, respiratory gases can be carried, to a small extent, in the haemolymph of some species with open circulatory systems.

Due to their large size, multicellular animals have evolved exchange surfaces, which you learned about in Chapter 7, and internal transport systems, which distribute nutrients and oxygen around their bodies. Here we discuss the different transport systems that have evolved in animals.

Why are transport systems needed?

The rate at which substances can travel through multicellular animals by diffusion alone is too slow to meet their requirements. The evolution of internal transport systems enables substances to be circulated at a faster rate. Specialised transport systems are necessary because multicellular animals have:

- high metabolic rates (and high demand for oxygen and nutrients)
- relatively small surface area to volume ratios
- molecules (e.g. enzymes and hormones) that need to be transported to specific tissues.

Transport systems in animals

Internal transport (circulatory) systems can be labelled as either open or closed.

▼ **Table 1** *A comparison of open and closed circulatory systems*

	Open circulatory system	Closed circulatory system
Description	Few vessels; the transport medium (haemolymph) is pumped from the heart into the body cavity (haemocoel)	Transport medium (blood) is enclosed in vessels
Examples	Some invertebrates, such as arthropods (including insects) and molluscs	All vertebrates, and many invertebrates (e.g. cephalopods and annelid worms)

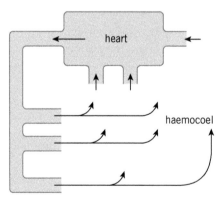

▲ **Figure 1** *A generalised open circulatory system*

Closed circulatory systems vary in their efficiency. Some species have a single circulatory system; others have evolved a double system, which enables a faster blood flow to be maintained. Species with double circulatory systems are able to sustain higher metabolic rates.

▼ **Table 2** *A comparison of single and double circulatory systems*

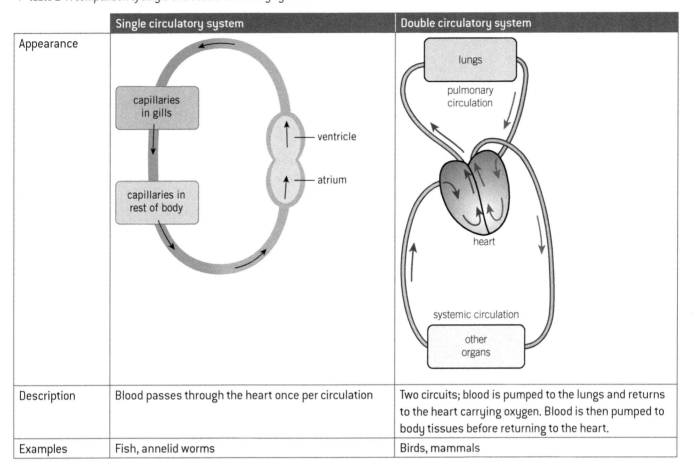

	Single circulatory system	Double circulatory system
Appearance	capillaries in gills · capillaries in rest of body · ventricle · atrium	lungs · pulmonary circulation · heart · systemic circulation · other organs
Description	Blood passes through the heart once per circulation	Two circuits; blood is pumped to the lungs and returns to the heart carrying oxygen. Blood is then pumped to body tissues before returning to the heart.
Examples	Fish, annelid worms	Birds, mammals

Summary questions

1 Flatworms lack a specialised circulatory system. As their name suggests, these animals have flattened body shapes. State and explain how nutrients and respiratory gases are transported in flatworms. *(2 marks)*

2 Explain the advantages of a closed double circulatory system over an open single circulatory system. *(3 marks)*

3 Fish possess closed single circulatory systems, but they are able to maintain high activity levels compared to other species with this type of system. Suggest why fish can be relatively active despite their single circulatory systems. *(3 marks)*

8.2 Blood vessels

Specification reference: 3.1.2(c)

You were introduced to the idea of closed circulatory systems in Topic 8.1, Transport systems in multicellular animals. Circulatory systems in animals include a variety of vessels, which are the subject of this topic.

The structures and functions of blood vessels

Blood travels through arteries after leaving the heart. Arteries branch into smaller arterioles, which diverge into capillaries. As blood travels back to the heart, capillaries converge to form venules, which lead into larger veins. The vessels' structural adaptations are suited to their positions in the circulatory system.

Arteries and veins consist of four layers: an inner **endothelium** layer, **elastic fibres**, **smooth muscle**, and **collagen** fibres. The proportions of each tissue present in vessels vary, depending on the requirements of the vessel. Capillaries, in contrast, have a single layer of endothelium cells.

Synoptic link

You learned about the structure of collagen and elastin in Topic 3.6, Structure of proteins.

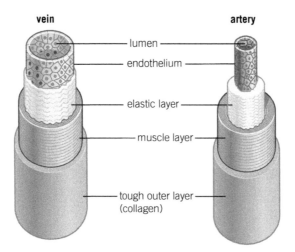

▲ **Figure 1** *Arteries and veins contain the same four tissues, but veins have wider lumens and less elastic tissue than arteries*

Revision tip: Blood vessel layers

The roles of the three outer layers in blood vessel walls can be summarised as follows:

Elastic fibres: provides flexibility

Smooth muscle: adjusts lumen size

Collagen: structural support

▼ **Table 1** *A comparison of blood vessel structure and function*

Vessel	Typical diameter	Relative proportion of elastin smooth muscle collagen	Key features	Explanation of key features
Arteries	5000 µm (0.5 cm)		High proportion of elastic tissue	To stretch and recoil, which prevents rupture when the heart pumps
Arterioles	50 µm		Smooth muscle	The muscle contracts to narrow the lumen (vasoconstriction) and relaxes to widen the lumen (vasodilation). This controls where blood flows
Capillaries	10 µm	None	Thin walls (single layer of flattened endothelial cells, with gaps between them)	Permeable wall to enable diffusion of particles into tissue fluid (see Topic 8.3, Blood, tissue fluid, and lymph)
Venules	100 µm	Very little elastin or smooth muscle	Thin walls (compared to veins)	Some permeability is retained, allowing the continued diffusion of some particles across the wall
Veins	10 000 µm (1 cm)		Wide lumen Valves, in most veins	Smooth blood flow at low pressure Backflow of blood is prevented

Summary questions

1 Explain how the structure of capillaries is suited to their function.

(3 marks)

2 Explain the difference in elastic fibre content between arteries and veins. *(2 marks)*

3 Medium-sized veins can have diameters of 1 cm, whereas medium-sized arteries have diameters of approximately 0.5 cm. **a** Express these measurements in standard form. **b** Explain the difference in the typical diameters of arteries and veins. *(3 marks)*

8.3 Blood, tissue fluid, and lymph

Specification reference: 3.1.2(d)

Synoptic link

You learned about osmosis in Topic 5.5, Osmosis.

In Topic 8.2, Blood vessels, we examined how blood is transported in vessels. Blood delivers nutrients, oxygen, and chemical messengers to tissues, and removes waste products, such as CO_2. Yet this is only part of the story. Two other transport fluids exist in animals: tissue fluid and lymph. Here you will learn about the composition and roles of all three fluids.

Tissue fluid formation

Tissue fluid is formed when water diffuses out of blood capillaries, carrying dissolved solutes across the capillary wall. The key to tissue fluid formation is the balance between two pressures: osmotic and hydrostatic.

Along the length of a capillary, water potential is always higher in tissue fluid (and osmotic pressure is therefore higher in the blood). However, hydrostatic pressure changes along the length of a capillary.

- *At the arterial end of a capillary*: the hydrostatic pressure of the blood is high. This outweighs the higher water potential in the tissue fluid. Water diffuses out of the capillary.

- *At the venous end of the capillary*: the hydrostatic pressure of the blood is too low to outweigh the higher water potential in the tissue fluid. Water diffuses back into the capillary.

Cells are bathed in tissue fluid. This enables an exchange of materials: oxygen and nutrients enter cells, and waste products leave cells. Molecules such as hormones can also move between the tissue fluid and cells.

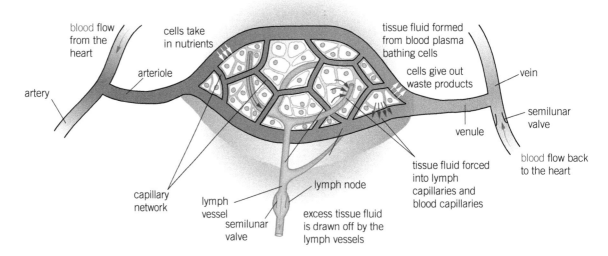

▲ **Figure 1** *The formation of tissue fluid*

Common misconception: Two pressures … but can we consider one to be a 'pull'?

The direction that water moves though capillary walls is determined by the balance of two types of pressure: hydrostatic and osmotic.

Hydrostatic pressure = the pressure exerted by a fluid in confined spaces.

Osmotic pressure is more difficult to visualise. Water moves from an area of high water potential (with less dissolved solute) to an area of low water potential (with more solute). The solution with the lower water potential has the higher osmotic pressure; it will attract water via osmosis.

Considering osmotic pressure as a 'pull' is perhaps easier; osmotic pressure rises as solute concentration rises, which increases the probability of a solution attracting water.

Blood plasma in capillaries exerts a greater osmotic pressure than tissue fluid because it has more solute dissolved. In this case, proteins in blood (e.g. albumin) contribute significantly to the osmotic pressure. This represents a specific form of osmotic pressure called **oncotic pressure**.

Lymph

Some tissue fluid (approximately 10%) drains into the lymphatic system rather than re-entering the blood. Lymph fluid passes through lymph vessels, via lymph nodes, before returning to the bloodstream.

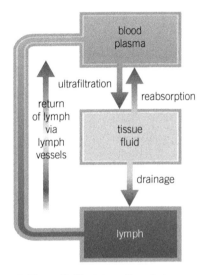

▲ **Figure 2** *The interactions between the three transport fluids*

Synoptic link

Topic 12.6, The specific immune system, provides details of the roles of leucocytes.

Functions and composition of the fluids

Fluid	Functions	Composition
Blood	Transport to and from tissues (the blood also plays a role in temperature regulation and as a pH buffer)	55% plasma (i.e. water plus solutes such as glucose, amino acids, ions, and large proteins, including hormones). 45% cells (leucocytes (white blood cells) and erythrocytes) and platelets.
Tissue fluid	Bathes cells, and exchanges materials with them	Few cells (which remain in the blood) other than some phagocytes (a type of leucocyte). Less solute than blood because many substances (e.g. oxygen, glucose, amino acids) diffuse into cells; no plasma proteins (e.g. albumin, fibrinogen).
Lymph	An important part of the immune system (e.g. phagocytes in lymph nodes ingest bacteria)	Similar to tissue fluid, but with less oxygen and nutrients, a greater proportion of fatty acids, and a large number of leucocytes.

Summary questions

1 State the differences between blood plasma and tissue fluid. (*3 marks*)

2 Explain the differences in the composition of tissue fluid and lymph. (*4 mark*)

3 Explain how tissue fluid is formed and reabsorbed into capillaries. (*5 marks*)

8.4 Transport of oxygen and carbon dioxide in the blood

Specification reference: 3.1.2(i) and (j)

Synoptic link

You read an overview of erythrocyte structure and function in Topic 6.4, The organisation and specialisation of cells. In Topic 3.7, Types of proteins, you learned about the structure of haemoglobin.

One of the functions of blood, as the transport medium in animals, is to deliver oxygen to respiring tissues and take away carbon dioxide. Here we examine how the transport and transfer of these respiratory gases is achieved in humans and other vertebrates.

Transport of gases in the blood

Oxygen transport

Oxygen binds to haemoglobin in red blood cells (erythrocytes). 98% of oxygen in the blood is transported by haemoglobin; 2% is carried in solution in the plasma. The reaction between haemoglobin and oxygen is reversible, which is crucial because O_2 is therefore released under the right conditions (see 'The Bohr effect', below).

$$\begin{array}{ccccc} Hb & + & 4O_2 & \rightleftharpoons & Hb(O_2)_4 \\ \text{Haemoglobin} & + & \text{oxygen} & \rightleftharpoons & \text{oxyhaemoglobin} \end{array}$$

Revision tip: Cooperation is the key

One haemoglobin molecule can bind with four O_2 molecules. One model of haemoglobin's chemistry suggests that the binding of the first O_2 molecule causes haemoglobin to change shape. This makes it easier for subsequent O_2 molecules to bind. This is known as positive cooperative binding.

Carbon dioxide transport

Small amounts of CO_2 are dissolved in blood plasma (5%) or combined with haemoglobin (10–20%). However, the majority (75–85%) is converted to HCO_3^- (hydrogen carbonate) ions.

$$CO_2 + H_2O \rightleftharpoons H_2CO_3 \rightleftharpoons HCO_3^- + H^+$$

Revision tip: Why is HCO_3^- ion production important?

The removal of carbon dioxide from plasma (and its conversion into HCO_3^- ions) maintains a steep CO_2 concentration gradient between respiring tissues and blood.

Chloride shift

The formation of HCO_3^- occurs in erythrocytes. HCO_3^- ions then move out of erythrocytes and into plasma. Chloride ions move into erythrocytes to replace them; this exchange is called the chloride shift.

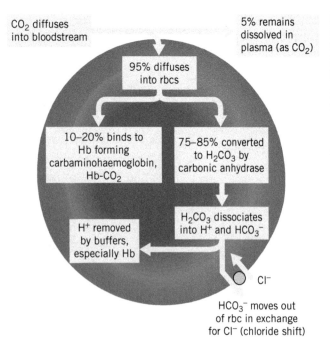

CO₂ diffuses into bloodstream

5% remains dissolved in plasma (as CO₂)

95% diffuses into rbcs

10–20% binds to Hb forming carbaminohaemoglobin, Hb-CO₂

75–85% converted to H₂CO₃ by carbonic anhydrase

H₂CO₃ dissociates into H⁺ and HCO₃⁻

H⁺ removed by buffers, especially Hb

Cl⁻

HCO₃⁻ moves out of rbc in exchange for Cl⁻ (chloride shift)

▲ **Figure 1** *Conversion of CO_2 to HCO_3^- ions is followed by the chloride shift (which transports HCO_3^- into the plasma)*

Oxygen dissociation curves

Oxygen dissociation curves illustrate how partial pressure affects the amount of oxygen binding to haemoglobin (i.e. % saturation). Small changes in oxygen partial pressure have a large influence on haemoglobin saturation.

The Bohr effect

Oxygen is released from haemoglobin more easily when carbon dioxide partial pressure (concentration) increases. This is known as the Bohr effect. CO_2 is produced in respiring tissues. This means O_2 will tend to be released where it is required, within respiring tissues. Conversely, CO_2 concentration in the lungs is low, which means O_2 binds more readily to haemoglobin in pulmonary capillaries.

Fetal haemoglobin

Fetal haemoglobin has a greater affinity for oxygen than adult haemoglobin (i.e. at a given partial pressure, fetal haemoglobin will be more saturated with oxygen than adult haemoglobin). This is important because it enables oxygen to be transferred from a mother's haemoglobin to a fetus's haemoglobin in the placenta.

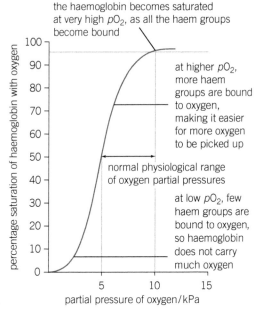

the haemoglobin becomes saturated at very high pO_2, as all the haem groups become bound

at higher pO_2, more haem groups are bound to oxygen, making it easier for more oxygen to be picked up

normal physiological range of oxygen partial pressures

at low pO_2, few haem groups are bound to oxygen, so haemoglobin does not carry much oxygen

▲ **Figure 2** *The oxygen dissociation curve for adult haemoglobin*

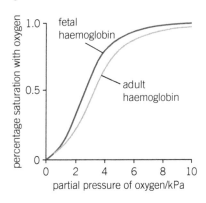

▲ **Figure 3** *Fetal haemoglobin binds to more oxygen than adult haemoglobin at a given partial pressure*

Revision tip: Partial pressures

The pressure exerted by a particular gas in a mixture is known as its partial pressure. In many circumstances, you can think of partial pressure as being equivalent to concentration.

Summary questions

1   **a** Describe the Bohr effect.
    b Sketch a graph to show two oxygen dissociation curves: one at high CO_2 partial pressure and one at low CO_2 partial pressure.
    (3 marks)

2   Fetal blood exiting the placenta has an O_2 partial pressure of approximately 3 kPa. Use the graph in Figure 3 to estimate the difference in oxygen saturation between fetal and adult haemoglobin at 3 kPa. *(1 mark)*

3   Describe the shape of the oxygen dissociation curve for adult haemoglobin and explain the significance of the shape for human physiology. *(3 marks)*

➕ **Go further: Other oxygen-carrying proteins are available …**

Haemoglobin is the molecule to which oxygen binds in the blood of vertebrates. Many **variants of haemoglobin** exist in humans and between species. Salt water crocodiles (*Crocodylus porosus*), for example, have been shown to produce haemoglobin with a different structure to haemoglobin in mammals and birds.

1   Mammals and birds are endotherms (animals that generate their own heat to maintain body temperature), whereas crocodiles are not. Suggest how the haemoglobin of crocodiles differs from that of endothermic species.

    Myoglobin is an oxygen-binding protein found in vertebrate muscles. It has a higher affinity than haemoglobin for O_2, except at very high oxygen partial pressures. Myoglobin releases O_2 at very low oxygen partial pressure.

2   Suggest myoglobin's function in muscles, and sketch oxygen dissociation curves for myoglobin and haemoglobin on the same graph.

    Oxygen is transported through some invertebrates by **haemocyanin**.

3   Haemocyanin is blue when oxygenated. Suggest a reason for the colour difference between haemoglobin and haemocyanin.

8.5 The heart

Specification reference: 3.1.2(e), (f), (g), and (h)

The heart is the organ that pumps blood around an organism. Hearts range in complexity from a basic muscular tube in invertebrates to the four-chambered organ found in mammals. We focus on the mammalian heart in this topic.

Heart structure

The mammalian heart consists of four chambers. The right and left sides fill and empty together; they are divided by a septum.

▼ **Table 1** *The movement of blood in and out of the heart's chambers*

Heart chamber	Where does blood flow from when it enters the chamber?	Where does blood flow to when it leaves the chamber?
Right atrium	Deoxygenated blood from the vena cava	Right ventricle
Right ventricle	Right atrium (through atrioventricular valve)	Pulmonary arteries (through semilunar valve)
Left atrium	Oxygenated blood from pulmonary veins	Left ventricle
Left ventricle	Left atrium (through atrioventricular valve)	Aorta (through semilunar valve)

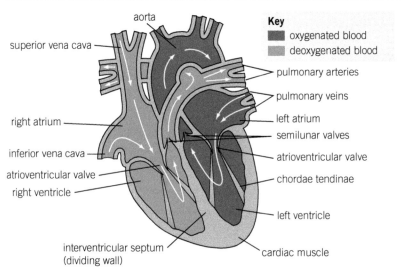

▲ **Figure 1** *The internal structure of the heart*

The cardiac cycle

The cardiac cycle comprises the events during one heartbeat. The cycle can be visualised in three stages: diastole (relaxation), atrial systole (the atria contract), and ventricular systole (the ventricles contract), which are illustrated below.

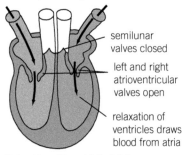

1. blood enters atria and ventricles from pulmonary veins and vena cava

semilunar valves closed

left and right atrioventricular valves open

relaxation of ventricles draws blood from atria

Relaxation of heart (diastole)
Atria are relaxed and fill with blood. Ventricles are also relaxed.

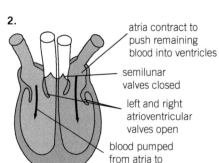

2. atria contract to push remaining blood into ventricles

semilunar valves closed

left and right atrioventricular valves open

blood pumped from atria to ventricles

Contraction of atria (atrial systole)
Atria contract, pushing blood into the ventricles. Ventricles remain relaxed.

3. blood pumped into pulmonary arteries and the aorta

semilunar valves open

left and right atrioventricular valves closed

ventricles contract

Contraction of ventricles (ventricular systole)
Atria relax. Ventricles contract, pushing blood away from heart through pulmonary arteries and the aorta.

▲ **Figure 2** *The cardiac cycle*

What controls the rhythm of the cardiac cycle?

The cycle is coordinated through the following steps:

- The sino-atrial node (**SAN**), which is the heart's pacemaker, initiates a wave of electrical excitation.
- The atria are stimulated to contract. (*This is the atrial systole.*)
- The electrical impulse reaches the atrio-ventricular node (**AVN**).
- After a delay, the electrical activity passes down the **bundle of His** (made of Purkinje fibres).
- The ventricles contract from the apex (i.e. the bottom of the heart in most diagrams). (*This is the ventricular systole.*)

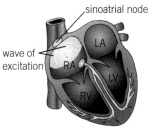

a wave of electrical activity spreads out from the sinoatrial node

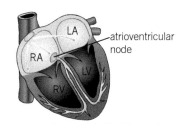

b wave spreads across both atria causing them to contract and reaches the atrioventricular node

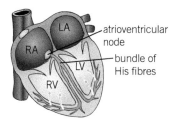

c atrioventricular node conveys a wave of electrical activity between the ventricles along the bundle of His and releases it at the apex, causing the ventricles to contract

▲ **Figure 3** *Control of the cardiac cycle*

Electrocardiograms (ECGs)

Electrocardiography can be used to monitor the electrical activity of a heart to check that a person's cardiac cycle is healthy. The results are shown on an ECG.

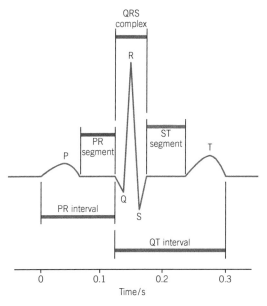

▲ **Figure 4** *An ECG trace for a single cardiac cycle. P corresponds to atrial systole; QRS represents ventricular systole; T is associated with diastole*

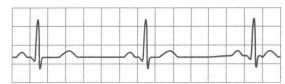

(b) Bradycardia – slow heart rate – beats evenly spaced, rate <60/min

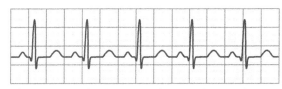

(c) Tachycardia – fast heart rate – beats evenly spaced, rate >100/min

▲ **Figure 5** *Two ECG traces showing abnormal heart rates*

Summary questions

1 Describe what occurs during **a** diastole **b** atrial systole **c** ventricular systole. *(3 marks)*

2 Explain why there is a delay between the P and QRS waves on an ECG trace. *(2 marks)*

3 Suggest what changes in atrial and ventricular pressure will occur through the cardiac cycle. *(3 marks)*

1 Which of the following statements is true of double circulatory systems?

    A They are open systems.

    B They are found in some invertebrate species.

    C They use haemolymph as a transport medium.

    D They are found in bird species. *(1 mark)*

2 Which of the following statements is true of arteries?

    A They possess an inner endothelium layer.

    B Their lumens are always wider than the lumens of veins.

    C They possess an outer elastic layer.

    D Their elastic layers are narrower than those of veins. *(1 mark)*

3 Which of the following statements is true of lymph?

    A It contains a higher concentration of oxygen than blood plasma.

    B Lymph vessels contain valves.

    C It contains no fatty acids.

    D Lymph drains into tissue fluid. *(1 mark)*

4 Which of the following statements is true of carbon dioxide transport within the blood?

    A Approximately 30% of CO_2 is dissolved in blood plasma.

    B Carbon dioxide is converted to HCO_3^- ions in blood plasma.

    C More oxygen binds to haemoglobin at high partial pressures of carbon dioxide.

    D Approximately 15% of carbon dioxide in the blood binds to haemoglobin.

    (1 mark)

5 Which of the following terms describes a heart rate that is abnormally fast?

    A Diastole

    B Systole

    C Tachycardia

    D Bradycardia *(1 mark)*

6 The following passage describes the formation of tissue fluid. Complete the passage by choosing the most appropriate word to place in each gap.

Tissue fluid is formed when water in blood .... diffuses out of capillaries. .... pressure is higher at the arterial end of a capillary. Water .... remains higher in the tissue fluid along the length of a capillary. Water diffuses back into capillaries at the .... end.

    (4 marks)

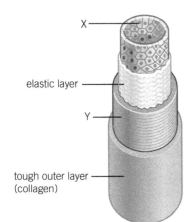

X —

elastic layer ——

Y —

tough outer layer —— (collagen)

7 The figure to the left shows the generalised structure of a vein.

    A Name the layers labelled X and Y. *(2 marks)*

    B Describe the role of the elastic layer within blood vessels. *(2 marks)*

9.1 Transport systems in dicotyledonous plants

Specification reference: 3.1.3(a) and (b)(i–ii)

In Chapter 8 you learned about the transport systems that have evolved in animals. Plants also require internal systems to transport water and nutrients between roots, stems, and leaves.

Why do plants need transport systems?

As with multicellular animals, relying on diffusion alone to transport molecules and ions would fail to meet the demands of plants. They have high metabolic rates, can grow to remarkable sizes, and certain parts of plants (e.g. the trunks of trees) have small surface area to volume ratios.

Vascular systems

Vertebrate animals transport water and sugars within the same vessels. Plants, in contrast, employ separate vessels to carry water and minerals (**xylem** vessels) and sugars (**phloem**). These two tissues together are known as the vascular system. Vascular tissue has different arrangements in the roots, stems, and leaves of plants.

Revision tip: What is a dicotyledonous plant?
Dicotyledons are flowering plants with seeds that grow two primary leaves (cotyledons). As you may have guessed, monocotyledons (the other category of flowering plant) produce seeds that grow one primary leaf. Their vascular tissue has a different arrangement, which you are not required to learn.

Synoptic link

You can remind yourself about surface area to volume ratios in Topic 7.1, Specialised exchange surfaces. Topic 8.1, Transport systems in multicellular animals, outlines the circulatory systems that animals have evolved.

▼ **Table 1** *The arrangement of vascular tissue in different parts of a plant*

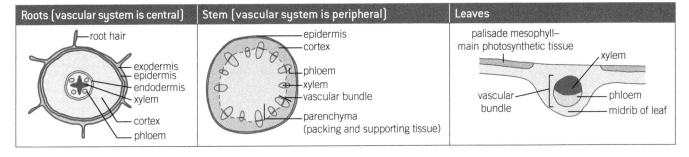

Roots (vascular system is central)	Stem (vascular system is peripheral)	Leaves
root hair; exodermis; epidermis; endodermis; xylem; cortex; phloem	epidermis; cortex; phloem; xylem; vascular bundle; parenchyma (packing and supporting tissue)	palisade mesophyll– main photosynthetic tissue; xylem; vascular bundle; phloem; midrib of leaf

Xylem and phloem

▼ **Table 2** *The structure and function of xylem and phloem tissues*

	Key features	Functions	Appearance
Xylem (see Topic 9.3, Transpiration, for more details of water movement up xylem vessels)	Dead cells have been fused to form hollow vessels, which are strengthened by lignin. Pits in the xylem wall enable water to move out into adjacent xylem vessels or other cells. Xylem fibres add strength, and xylem parenchyma cells store food	Transport of water and mineral ions up a plant. The tissue also provides structural support for the plant.	xylem parenchyma; lignified cell wall; lumen of xylem vessel

| Phloem (see Topic 9.4, Translocation) | Sieve tube elements are living cells joined end-to-end, forming a tube with internal pores (sieve plates)

Companion cells carry out all the metabolic functions of the phloem tissue (as the tubes lack nuclei). Materials pass into sieve tubes via plasmodesmata (cell wall channels). | Transport of solutes (e.g. sugars and amino acids) up and down a plant | 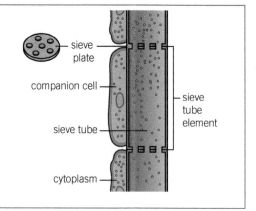 |

Revision tip: Dead or alive?

Xylem vessels and phloem sieve tubes both lack nuclei, but phloem sieve tubes are made from living cells, unlike xylem vessels.

Practical skill: examining plant tissue

Xylem vessels can be observed in either dead or living tissue. Plant stems or roots can be cut into thin sections, stained, and prepared on slides. The tissue is observed under a light microscope.

Xylem in some plant species (e.g. celery or gerberas) can be observed without the need to cut away tissue. The plants are first soaked in a solution of dye for at least 24 hours.

Summary questions

1 Describe the different arrangement of vascular tissue in roots and stems. *(2 marks)*

2 Outline how xylem vessels are adapted for water transport. *(3 marks)*

3 Explain how sieve tube elements survive, despite lacking nuclei and containing a small amount of cytoplasm. *(2 marks)*

9.2 Water transport in multicellular plants

Specification reference: 3.1.3(d)

Water plays numerous vital roles in plants, such as maintaining turgor, as a transport medium for mineral ions and sugars, and as a raw material in photosynthesis. Without a heart to pump fluid through its vessels, however, a plant relies on alternative mechanisms to transport water. You will learn about these mechanisms in the following two topics. We begin by looking at water movement from the soil to the xylem vessels.

Water uptake

Root hair cells are adapted to take up water from the surrounding soil because they are:

- long and narrow, which increases surface area to volume ratio
- able to penetrate between soil particles
- able to maintain a water potential gradient between the soil and the cell (due to solutes dissolved in the root).

Water moves into root hair cells by osmosis.

Movement towards the xylem

Water moves from cell to cell towards the xylem either through the cytoplasm (the **symplast** pathway) or through cell walls (the **apoplast** pathway). The flow of water is maintained by the *transpiration pull*, which you will learn about in Topic 9.3, Transpiration.

Entering the xylem

Water reaches the layer of cells (the **endodermis**) surrounding xylem vessels. An impermeable band (the **Casparian strip**) around endodermal cells forces water in the apoplast pathway back into the cytoplasm.

Endodermal cells move mineral ions into the xylem by active transport. As a consequence, the water potential in endodermal cells is higher than that of the xylem. Water diffuses into the xylem by osmosis. The initial flow into vascular tissue helps to force water up a stem; this is known as **root pressure**. The transpiration pull, which we discuss in the next topic, is a more significant factor in water movement than root pressure.

Question and model answer: Assessing evidence

Q. Outline the experimental evidence that supports the role of active transport in producing root pressure.

A. Active transport requires ATP, which is reliant on oxygen and carbohydrates (respiratory substrates). Root pressure drops when the concentrations of oxygen or carbohydrates are decreased. Cyanide prevents the production of ATP in respiration. The application of cyanide to root cells makes root pressure disappear. Root pressure increases as temperature is increased – this indicates that chemical reactions are likely to be occurring.

Summary questions

1. State two uses of water in plants. *(2 marks)*

2. Describe and explain the function of the Casparian strip. *(3 marks)*

3. Compare and contrast the movement of water through the apoplast and symplast pathways. *(5 marks)*

Revision tip: Water uptake – what's the solution?

The cytoplasm of root hair cells contains solutes such as sucrose and amino acids. Active transport of mineral ions from the soil further decreases the water potential inside the roots. This enables water to diffuse by osmosis down a water potential gradient from the soil.

Synoptic links

You read about the adaptations of root hair cells in Topic 6.4, The organisation and specialisation of cells.

The principles of water diffusion are outlined in Topic 5.5, Osmosis.

Revision tip: The Casparian strip

The Casparian strip is waterproof because it is composed largely of a waxy substance called suberin. The strip prevents toxic solutes from continuing to move into the plant, and it stops water from returning to the root cortex from xylem vessels.

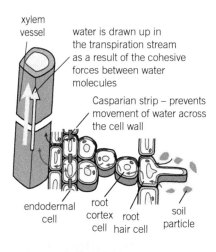

▲ **Figure 1** *Water movement from the soil to xylem tissue*

9.3 Transpiration

Specification reference: 3.1.3(c) and (d)

Synoptic link

The properties of water were discussed in Topic 3.2, Water.

Key term

Transpiration: The evaporation of water from a plant's leaves. (Water evaporates from cells inside the leaves and diffuses out of the stomata.)

Revision tip: A sticky subject

Try to avoid confusing *cohesion* (the attraction between water molecules) and *adhesion* (the attraction of water molecules to the inner surface of xylem vessels).

Synoptic link

You learned about the structure of guard cells in Topic 6.4, The organisation and specialisation of cells.

We examined how water enters plant roots in Topic 9.2, Water transport in multicellular plants. Here you will learn how water is moved through the rest of a plant.

Transpiration and water movement through xylem

- Water leaves a plant by **transpiration** (principally though stomata, which are discussed below).
- Water is pulled up through xylem vessels to replace the water lost through transpiration (i.e. the **transpiration pull**).
- Water molecules cohere to each other (i.e. they are attracted to each other through hydrogen bonding; they exhibit **cohesion**). This enables an unbroken chain of water molecules to be pulled up the xylem vessels (i.e. the **transpiration stream**).
- In addition, water molecules adhere to the sides of xylem vessels, which helps move the transpiration stream up the narrow vessels. This is called **capillary action**.

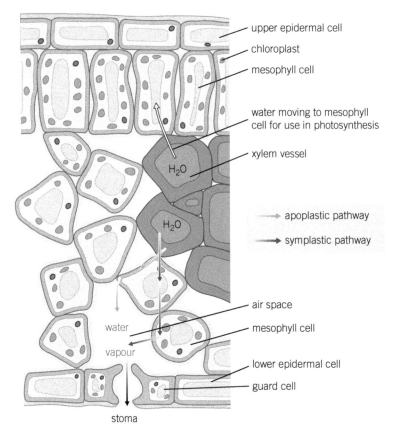

▲ **Figure 1** *Water movement within a leaf*

Evidence for the cohesion–tension theory

The model of water movement from soil to the leaves is known as the **cohesion–tension theory**. Much evidence has been gathered to support the theory, including:

- Trees become narrower when they transpire; this can be explained by the increased tension in xylem vessels during high rates of transpiration.
- Air is sucked up (rather than water leaking out) when a stem is cut.

- Water is no longer moved up a broken stem because the air that is pulled in breaks the transpiration stream (i.e. there is no longer a continuous chain of water molecules).

Stomata

In leaves, some water is used in photosynthesis, some exits through stomata (singular: stoma). The opening and closing of stomata is controlled by guard cells. When open, stomata allow the exchange of carbon dioxide and oxygen, but this also leads to transpiration and the loss of water.

Factors affecting transpiration

Both environmental factors and factors within a plant will affect transpiration rate, as shown in the following table.

Factor	How does it affect transpiration?	What increases transpiration rate?
Light intensity	Stomata open in the light	Higher light intensity
Temperature	Changes the kinetic energy of molecules	Higher temperature
Humidity (of the air)	Affects the water potential gradient between leaf and air	Lower humidity
Air movement	Affects how quickly moist air is removed	More air movement
Number of leaves	Affects the surface area available for loss of water vapour	More leaves
Number of stomata	Alters how much water is able to diffuse from the leaves	More (and larger) stomata
Thickness of cuticle	Waxy cuticles reduce water loss	Thin (or no) cuticles

Practical skill: measuring transpiration

Measuring water loss from plants is problematic. However, plants transpire approximately 99% of the water they absorb from the soil. Therefore measuring water uptake can give a good estimate of transpiration rates. Water uptake tends to be measured with a **potometer**.

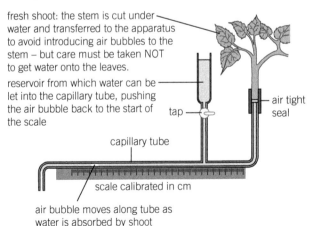

fresh shoot: the stem is cut under water and transferred to the apparatus to avoid introducing air bubbles to the stem – but care must be taken NOT to get water onto the leaves.

reservoir from which water can be let into the capillary tube, pushing the air bubble back to the start of the scale

tap

air tight seal

capillary tube

scale calibrated in cm

air bubble moves along tube as water is absorbed by shoot

▲ **Figure 2** *A potometer*

The rate of water uptake is calculated by measuring the distance moved by the air bubble after a set time. Independent variables (e.g. temperature or light intensity) can be tested using a single plant species. Alternatively, water uptake in different species can be compared, which requires all environmental factors to be controlled (i.e. a constant temperature and light intensity).

Summary questions

1 State how each of the following conditions would affect transpiration rate. Explain your answers.
   a Decreased temperature
   b Increased light intensity
   c Absence of waxy cuticle.
   (6 marks)

2 Explain why the number of stomata in a leaf is likely to represent a compromise that depends on the environmental conditions in a plant's habitat. *(2 marks)*

3 The volume of water taken up by a potometer is calculated using this formula:
   Volume = distance moved by bubble $\times \pi r^2$ (where r = the radius of the capillary tube in the potometer)
   If the mean distance moved by an air bubble is 12.0 mm in one minute, and the capillary tube has a diameter of 1.0 mm, calculate the rate of water uptake in $mm^3\,hr^{-1}$.
   (4 marks)

9.4 Translocation

Key term

Translocation: The movement of organic solutes through phloem sieve tubes.

Revision tip: Why transport sucrose?

Converting glucose to sucrose for translocation (and converting back to glucose at sinks) might seem inefficient. Glucose, however, is much more reactive than sucrose. This conversion prevents unwanted (and potentially harmful) reactions occurring during translocation.

Glucose produced in photosynthesis is converted to sucrose, which is transported around a plant. The transport of sucrose and other organic compounds occurs in phloem sieve tubes and is known as translocation.

Sources and sinks

Assimilates (the products of photosynthesis, such as sucrose) are translocated from sources to sinks. **Sources** include: green leaves and stems, and storage sites (such as tubers). **Sinks** include: developing seeds and fruits (which are laying down food stores), and growing roots.

The process of translocation

The movement of molecules into sieve tubes (phloem loading) can be an active or passive process (see Topic 9.2, Water transport in multicellular plants).

Passive loading: sucrose diffuses through cytoplasm and plasmodesmata (the symplast route) and enters sieve tubes.

Active loading: sucrose travels through cell walls and intercellular spaces (the apoplast route) and eventually reaches companion cells. The sucrose is loaded into companion cells through a combination of active transport and facilitated diffusion (as shown in Figure 1).

In both cases, once sucrose is in sieve elements, water enters the phloem by osmosis. Turgor (water) pressure causes movement (mass flow) of water and dissolved solutes to regions of lower pressure. Water and solutes are unloaded from the sieve tubes at sinks (either by active or passive processes).

The evidence

Evidence for the mechanism of translocation includes: observation of the proteins required for active transport (thanks to advances in microscopy) the lack of translocation if mitochondria in companion cells are poisoned, and analysis of flow rates using aphid stylets.

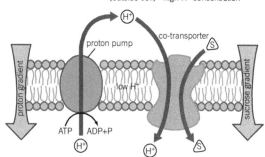
[outside cell] - high H⁺ concentration
[inside cell] - low H⁺ concentration

▲ **Figure 1** *The movement of sucrose into companion cells requires active transport*

Summary questions

1 State two examples of a translocation source and two examples of a sink. *(4 marks)*

2 Explain how sucrose is loaded into phloem cells using a combination of active transport and facilitated diffusion. *(4 marks)*

3 Suggest explanations for the distribution of carbohydrates in the parts of the plant shown in the following table. *(6 marks)*

Part of plant	Mean carbohydrate content ($\mu g\,g^{-1}$ fresh mass)			
	Sucrose	Glucose	Fructose	Starch
Leaf blade	1312	210	494	62
Vascular bundle in leaf stalk	5757	479	1303	<18
Tissue surrounding the vascular bundle in the leaf	417	624	1236	<18
Buds, roots, and tubers	2260	120	370	152

9.5 Plant adaptations to water availability

Specification reference: 3.1.3(e)

So far in this chapter you have examined how plants obtain, move, and use water. Some species, however, inhabit ecosystems in which water is scarce (**xerophytes**) or abundant (**hydrophytes**). Here you will learn about the adaptations evolved by both types of plant.

Xerophyte adaptations

Adaptation for conserving water	Benefit
Sunken stomata (in pits)	Traps moist air and reduces transpiration rate
Reduced number of stomata	Lowers the rate of transpiration
Reduction of leaf area (e.g. needle-like leaves in conifers)	Water loss via transpiration is reduced
Thick waxy cuticle	Reduces transpiration via the cuticle (which represents approximately 10% of water loss)
Curled leaves (e.g. in marram grass)	Traps moist air and reduces transpiration rate
Increased water storage (e.g. succulents such as cacti and samphire)	Provides a long-term reserve of water
Leaf loss during dry periods	Significantly reduces transpiration when no water is available
Long roots (e.g. some cactus' species)	Increases the chances of obtaining water from the ground

Hydrophyte adaptations

Adaptation for inhabiting aquatic ecosystems	Benefit
No waxy cuticle (or at least a very thin one)	The cuticle serves little purpose in hydrophytes because water loss is not an issue
Stomata tend to be open, and in plants with floating leaves stomata are on the upper surface	Gas exchange is maximised
A reduction in internal structural support	This would be another unnecessary feature because water can support aquatic plants
Air sacs and aerenchyma (specialised parenchyma tissue containing air spaces)	Increased buoyancy; leaves and flowers can float on the water surface
Small roots	Water uptake via roots is less significant because water can diffuse directly into other parts of submerged plants

Key term

Adaptation: A trait that benefits an organism in its environment and increases its chances of survival and reproduction.

Synoptic link

You will learn more about the adaptations of marram grass in Topic 10.7, Adaptations.

Synoptic link

You read about stomata in Topic 6.4, The organisation and specialisation of cells, and Topic 9.3, Transpiration.

Summary questions

1   List three ways in which xerophytes can be adapted to ecosystems with a lack of water. *(3 marks)*

2   Suggest how the stomata in hydrophytes might differ from those in land-based plants. Explain your answer. *(2 marks)*

3   Suggest why the results from a potometer experiment are not entirely representative of the transpiration rate of a plant in its natural habitat. *(3 marks)*

1   The volume of water taken up by a plant in a potometer is calculated using this formula: volume = length of air bubble $\times \pi r^2$ (where r = radius of the capillary tube).

In an experiment, the mean distance moved by the air bubble in a potometer was 1.5 cm in one minute. The capillary tube had a diameter of 1.0 mm. Which of the following is the rate of water uptake in $mm^3 hr^{-1}$?

A 11.8   C 706.5

B 70.7   D 2826.0   *(1 mark)*

2   Which of the following statements is/are true of the possible adaptations exhibited by hydrophytes?

1   Air spaces within stems

2   Rolled leaves to trap water vapour

3   Needle-shaped leaves with reduced surface areas

A 1, 2, and 3 are correct

B Only 1 and 2 are correct

C Only 2 and 3 are correct

D Only 1 is correct   *(1 mark)*

3   Which of the following does water pass through in the apoplast pathway?

A Plasmodesmata   C Vacuoles

B Cell walls   D Cytoplasm   *(1 mark)*

4   Which of the following statements is true of carbohydrate loading into companion cells?

A Sucrose is actively transported.

B Glucose is actively transported.

C Sucrose is co-transported.

D Glucose is co-transported.   *(1 mark)*

5   Outline the evidence for the role of active transport in maintaining root pressure.   *(3 marks)*

6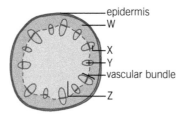

The figure shows the stem of a herbaceous plant.

Name the tissues found at W, X, Y, and Z.   *(4 marks)*

7   Complete the following passage about the transpiration stream; choose the most appropriate word to place in each gap.

Water evaporates from .... cells in the leaves of plants. Water molecules diffuse out of stomata, down a concentration gradient. This lowers the water potential of cells within leaves. Water molecules move up .... Hydrogen bonds enable .... (i.e. water molecules being attracted to each other). Water moves up a plant's stem by .... action.

(4 marks)

10.1 Classification

Scientists estimate that more than 8 million species are alive on Earth today. Naming and grouping this vast array of organisms makes it easier for us to study them. The term we use for sorting organisms into groups is **classification**.

The classification system

Classification places organisms in taxonomic groups, or taxa (singular: taxon), which are organised in a hierarchy. The method scientists use to classify organisms is called the Linnaean system. Eight taxonomic levels are used, ranging from domains (the taxon containing the greatest number of organisms) to species (the smallest unit of classification). Table 1 illustrates the taxonomic groups of two species: common wheat and tigers.

▼ **Table 1** *The taxonomic groups of two eukaryotic species*

Taxonomic group	Organism	
	Common wheat	**Tiger**
Domain	Eukaryota	Eukaryota
Kingdom	Plantae	Animalia
Phylum	Magnoliophyta	Chordata
Class	Liliopsida	Mammalia
Order	Poales	Carnivora
Family	Poaceae	Felidae
Genus	*Triticum*	*Felix*
Species	*aestivum*	*tigris*

The number of species in each taxon increases as you move up the hierarchy. A genus (plural: genera), for example, contains at least one species, a family contains several genera, and an order comprises several families. The genetic similarity between organisms increases as you progress down the hierarchy from domain to species.

Advantages of a classification system

Classifying organisms into groups, using an agreed system, enables scientists to:

- analyse evolutionary relationships between organisms
- predict characteristics (as species grouped together are likely to share characteristics) and identify species
- share research findings (without confusion or ambiguity).

Key terms

Classification: The sorting of organisms into groups.

Species: A group of organisms that can interbreed to produce fertile offspring.

Revision tip: Mnemonics

Remembering the order of the taxonomic groups can be tricky. Inventing a mnemonic can help. One example is 'Don't Keep Pickled Cucumber Or Fried Gherkin Sauce'.

Revision tip: Naming species

The binomial system for naming species includes a few rules for you to remember:

1 The genus name begins with a capital letter.
2 The species name is always in lower case.
3 You should write both words in italics or, if handwritten, underline both words.

Summary questions

1 What is incorrect about each of the following binomial names?
   a *bufo fowleri* **b** Bufo americanus **c** *Turdus Merula* (*3 marks*)

2 Complete the Table 2 to show the classification of humans, *Homo sapiens*, and chimpanzees, *Pan troglodytes*. (*8 marks*)

3 The biological species concept defines a species as a group of organisms that can breed to produce fertile offspring. Suggest why this definition does not include all organisms and therefore may lack accuracy.(*2 marks*)

▼ **Table 2**

Taxon	Human	Chimpanzee
Domain		
Kingdom		
		Chordata
	Mammalia	Mammalia
	Primates	
	Hominidae	Hominidae
Genus		
Species		

10.2 The five kingdoms

Specification reference: 4.2.2(c)

Synoptic link

You learnt about the structure of prokaryotic and eukaryotic cells in Topic 2.4, Eukaryotic cell structure, and Topic 2.6, Prokaryotic and eukaryotic cells.

In Topic 10.1, Classification, you read about the principles of classification: how organisms are grouped together based on similarities. Here you will learn more about the taxonomic groups that have traditionally been considered the largest (the kingdoms). You will also consider a new level in the classification system (the domains) and the evidence that led to its development.

What are the five kingdoms?

The key features of the five kingdoms (Prokaryotae, Protoctista, Fungi, Plantae, and Animalia) are compared in Table 1.

▼ **Table 1** *A comparison of the features of the five kingdoms*

Kingdom	Unicellular or multicellular?	Domain(s)	Organelles present?	Cell wall	How do they gain nutrients?
Prokaryotae	unicellular	Bacteria and Archaea	No	Yes (peptidoglycan / murein)	autotrophic or heterotrophic
Protoctista	unicellular or multicellular	Eukarya	Yes	No	autotrophic or heterotrophic
Fungi	unicellular or multicellular	Eukarya	Yes	Yes (chitin)	saprotrophic (extracellular digestion)
Plantae	multicellular	Eukarya	Yes	Yes (cellulose)	autotrophic
Animalia	multicellular	Eukarya	Yes	No	heterotrophic

the three-domain system

Bacteria	Archaea	Eukarya

the six-kingdom system

Eubacteria	Archaebacteria	Protoctista	Fungi	Plantae	Animalia

the traditional five-kingdom system

Prokaryotae	Protoctista	Fungi	Plantae	Animalia

▲ **Figure 1** *The relationship between the three most common classification systems (three domains, six kingdoms, and five kingdoms)*

Why have domains been introduced?

A scientist called Carl Woese suggested dividing the Prokaryotae kingdom into two groups: Bacteria and Archaea. He placed the other four kingdoms (Protoctists, Fungi, Plants, and Animals) into a single group (the Eukarya, or Eukaryota). These three new groups were named domains.

Whereas early classification systems were based on observable anatomical features, modern systems rely on molecular comparisons between species. Woese's domain classification is based on several molecular observations, which include:

- Bacterial cell walls contain peptidoglycan but those of Archaea do not.
- The RNA polymerase of Archaea contains 8–10 subunits, but bacterial RNA polymerase contains only five subunits.
- Archaea have rRNA that is different to the rRNA of Bacteria and Eukarya.

Summary questions

1 State two differences between bacteria and fungi.   (2 marks)

2 Classify the following species into the correct kingdom and domain.
   a *Amoeba proteus* – a single-celled organism that lacks a cell wall.
   b *Serpula lacrimans* – a species that uses extracellular digestion to decompose organic material.
   c *Nanoarchaeum equitans* – a species discovered in 2002; their cell walls lack peptidoglycan, chitin, and cellulose.   (3 marks)

3 Explain why classification systems have changed over time, using the introduction of domains as an example.   (4 marks)

10.3 Phylogeny

Specification reference: 4.2.2(d)

In the previous topics you learned about how scientists classify organisms into groups that share particular features and evolutionary histories. The study of evolutionary relationships between species is called phylogeny. These relationships can be illustrated by constructing diagrams known as evolutionary (or phylogenetic) trees.

Key term

Phylogeny: The study of evolutionary relationships between organisms.

Producing phylogenetic trees

Whereas classification places species into discrete groups, phylogeny enables the construction of evolutionary trees, which show chronological relationships between species. These trees are not necessarily fixed. New evidence and different interpretations of evidence can result in the re-evaluation of evolutionary relationships. For example, a phylogenetic tree could be built from fossil evidence, but contradictory genetic evidence might indicate that an alternative phylogenetic tree is more appropriate.

Question and model answer: Interpreting phylogenetic trees

Q. Explain how the phylogenetic tree indicates which species is extinct.

A. Species E is extinct because the branch for this species ends before the present day.

Q. Which species is most closely related to species A? Explain your answer.

A. Species B, which shares a common ancestor with species A at node 1. This indicates they evolved into separate species relatively recently.

Q. What does node 4 in this phylogenetic tree represent?

A. Node 4 represents a common ancestor of species C, D, E, and F.

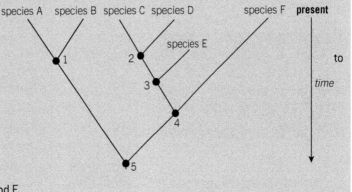

Summary questions

1 Explain why the structure of a phylogenetic tree may change over time. *(2 marks)*

2 The phylogenetic tree below shows one interpretation of the evolutionary relationships between apes.
   a Which was the first species in the tree to become extinct? *(1 mark)*
   b Are gorillas more closely related to humans or gibbons? *(1 mark)*

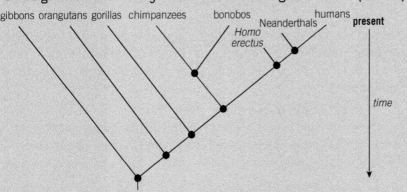

3 Suggest what advantages phylogenetic classification has over traditional hierarchical taxonomic classification. *(3 marks)*

Revision tip: Interpreting trees

You can interpret phylogenetic trees using the following rules:

- The earliest species is found at the base of the tree, and the most recent species are found at the tips of the branches.

- Branches for extinct species will end before the present day.

- The closer the relationship between two species, the more recently they will have branched from a common ancestor.

- Branch length is proportional to time.

- Nodes at branching points represent common ancestors.

10.4 Evidence for evolution

Phylogeny, which you studied in the previous topic, examines the evolutionary relationships between species. Here you will learn more about the theory of evolution, which explains how changes to species' characteristics occur over many generations.

Evidence for evolution

Evolution enables species to develop more advantageous phenotypes over time. These changes are caused by DNA mutations, which are inherited by subsequent generations. A wealth of evidence for evolution now exists, including the examples listed in the following table.

Type of evidence	What does the evidence show?
Palaeontology (fossils)	**Fossils** of simple organisms tend to be found in the oldest rocks, whereas more complex organisms are found in more recent rocks.
	Plant fossils appear in older rocks, before animals. This matches their ecological relationship (animals require plants to survive and will therefore have evolved later).
	Similarities between different fossil species (and extant species) reveal **gradual anatomical changes** over time.
Comparative anatomy	**Homologous structures** (anatomical features that have slight differences but the same underlying structure) provide evidence of divergent evolution (e.g. the vertebrate pentadactyl limb has evolved to perform a range of different functions in different species).
	The **embryos** of related species have similar appearances, despite considerable interspecific differences between adults. This indicates that different species have evolved from a common origin.
Comparative biochemistry	The rate of mutations in **DNA** can be calculated. This enables evolutionary relationships to be analysed. The closer the relationship between two species (i.e. the more recently their evolutionary paths diverged), the fewer the differences in their DNA base sequences (and therefore the fewer the differences in the primary structures of their proteins).

How was the theory of evolution developed?

Charles Darwin (in conjunction with Alfred Wallace) formed the initial theory of evolution by natural selection. Darwin was influenced by several ideas and observations, including:

- Charles Lyell's geological theories: fossils are evidence of animals living millions of years ago and natural processes can result from *gradual changes* and accumulations.
- His observations of finches on the Galapagos Islands: the slight differences (in beaks and claws) between species on neighbouring islands indicated that the birds were related but had developed different *adaptations* suited to their food sources.
- His studies of artificial selection by pigeon breeders enabled him to draw parallels with how *natural selection* might work. He concluded that *variation* exists between individuals in a population and the best adapted are more likely to survive and pass on their favourable characteristics to offspring.

Synoptic link

You will learn more about how natural selection works in Topic 10.8, Changing population characteristics.

Summary questions

1 What can the DNA base sequences of two species tell us about their evolutionary relationship? *(2 marks)*

2 Suggest why the fossil record fails to provide a complete picture of evolutionary history. *(3 marks)*

3 Explain the parallels between artificial selection and natural selection that Darwin observed. *(3 marks)*

10.5 Types of variation

Specification reference: 4.2.2(f)

In the previous topic we began to consider the principles of evolution: how species change over time. Evolution by natural selection can function only when there is variation between individuals in a population (intraspecific variation). The evolution of new species results in interspecific (between-species) variation. Here you will learn about the genetic and environmental causes of such variation.

The causes of variation

Variation between individuals can result from differences in their genetic material. Individuals of the same species can have different versions (alleles) of the same genes. Members of a species may have certain genes that other species lack. The environment introduces another layer of variation within populations. This is demonstrated by identical twins – they have the same DNA but will develop phenotypic variation during the course of their lifetimes based on differences in their environments.

Genetic variation

The following table outlines how genetic variation can be produced.

Cause of variation	How is variation introduced?	Synoptic link
DNA mutation	DNA base sequences are altered by mutations, producing new alleles.	2.1.3e Topic 3.9, DNA replication and the genetic code
Crossing over	Non-sister chromatids exchange genetic material during prophase I of meiosis, thereby creating new allele combinations.	2.1.6f Topic 6.3, Meiosis
Independent assortment	The random alignment of homologous chromosomes (metaphase I of meiosis) and sister chromatids (metaphase II) produces many allele combinations in gametes.	
Random fertilisation	During sexual reproduction, the identity of the two gametes that combine to form a zygote is largely due to chance.	

Only organisms that reproduce sexually experience crossing over, independent assortment, and random fertilisation. Mutations alone produce genetic variation in species that undergo asexual reproduction.

Environmental variation

Some traits are dictated solely by genetic variation (e.g. blood groups). Variation in characteristics is rarely determined by the environment alone, although physical damage (scarring) is one exception. Most variation results from a combination of genetics and the environment. The potential height to which a person can grow, for example, is governed by their genes, but their environment can influence whether or not they grow to this height (i.e. a poor diet may limit the person's height).

Revision tip: Intraspecific and interspecific variation

Variation between individuals in the same species is called in**tra**specific (you could remember this by thinking '**tra**pped within species'). This variation tends to involve minor differences (e.g. flower or fur colour, or height) and is a result of gene variants (i.e. members of a species have the same genes but can have different alleles of these genes).

Variation between members of different species is called in**ter**specific (*inter-* means 'between'; e.g. 'international' means 'between nations'). This variation tends to involve significant differences (although the closer the relationship between the species, the less significant the differences). Interspecific variation can be the result of species possessing different genes, as well as having different alleles.

Summary questions

1 State, with a reason, whether the following statements describe interspecific or intraspecific variation. **a** Arctic rose plants (*Rosa acicularis*) usually have pink flowers, but some have white flowers. **b** The shape of arctic rose (*Rosa acicularis*) leaves is pinnate. The desert rose (*Rosa stellata*) has trifoliate leaves. **c** Oystercatchers (*Haematopus ostralegus*) and zebra finches (*Taeniopygia guttata*) both have bills that vary in colour from orange to red. *(3 marks)*

2 Describe how genetic variation can be introduced into a species. *(4 marks)*

3 This graph shows some different conditions shared by twins and illustrates the relative influence of genes and the environment on each trait.
Explain the evidence for the comparative influence of genetics and the environment on height and strokes. *(3 marks)*

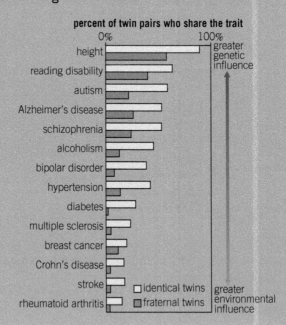

10.6 Representing variation graphically

Specification reference: 4.2.2(f)

We began thinking about biological variation in Topic 10.5, Types of variation. Variation within and between populations can be displayed and analysed by scientists in order to reveal patterns, make comparisons, and draw conclusions. Here you will learn about three statistical analysis techniques: standard deviation (for measuring the extent of variation in a population), Student's *t* test (for comparing data between two populations), and correlation coefficients (for analysing whether a correlation exists between two variables).

Discontinuous vs continuous variation

The type of data scientists collect determines how it should be represented graphically. Data can show either continuous or discontinuous variation. The features of both types of variation are outlined in the following table.

	Discontinuous variation	Continuous variation
Nature of the data	Discrete values with no intermediate values	A range of values (i.e. a continuum)
Genetic influence	One gene or a small number of genes	Several genes (polygenic)
Environmental influence	No	Yes
How is it represented?	Bar chart / pie chart	Histogram
Examples	Gender, blood groups, bacterial shape, whether flowering plants are monocots or dicots	Height, weight, number of pollen grains produced, rate of binary fission

Representing data for continuous variation

A trait that shows continuous variation often produces a normal distribution curve when population data are displayed in a graph. A **normal distribution curve** is a symmetrical, bell-shaped curve where most values lie close to the mean.

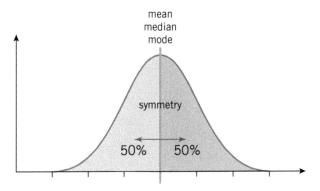

▲ **Figure 1** *A normal distribution curve*

Standard deviation

Standard deviation is calculated to assess the variation within a data set. A high standard deviation value indicates that the data shows a lot of variation. The calculation of standard deviation is performed on data with a normal distribution.

Worked example: Standard deviation

Standard deviation (σ) is calculated as:

$$\sigma = \sqrt{\frac{\sum (x - \bar{x})^2}{n-1}}$$

$\sum$ = the sum (total) of

x = value measured

$\bar{x}$ = mean value

n = total number of values in the sample

Here is a worked example of the calculation, using the horn length of topi antelopes as the data set:

Sample number	1	2	3	4	5	6	7	8	9	10
Horn length (cm)	38.4	39.0	39.2	40.1	39.4	38.9	40.2	39.2	40.8	40.8

1   First, calculate the mean $(\bar{x})$ by dividing the sum of the measurements by the number of individuals sampled. In this case, the calculation is 396 / 10 = 39.6 cm

2   The mean value is then subtracted from each of the 10 measured values for horn length $(x - \bar{x})$. For example, 38.4 − 39.6 = −1.2. This gives us a mixture of positive and negative values: −1.2, −0.6, −0.4, 0.5, −0.2, −0.7, 0.6, −0.4, 1.2, 1.2

3   These differences are then squared $(x - \bar{x})^2$ (e.g. $-1.2^2 = 1.44$) and the squared differences are added together $(\sum (x - \bar{x})^2)$. In this case, the calculation is 1.44 + 0.36 + 0.16 + 0.25 + 0.04 + 0.49 + 0.36 + 0.16 + 1.44 + 1.44 = 6.14 cm.

4   The sum of the squared differences $(\sum (x - \bar{x})^2)$ is divided by the sample size minus one. In this example, the calculation is 6.14/9 = 0.682.

5   We then find the square root of this value to calculate the standard deviation.
$\sqrt{0.682} = 0.826$

What does this σ value of 0.826 mean? It tells us that 68% of all measured values should be within 1 standard deviation of the mean. If we measured every topi in the population, we would expect 68% of individuals to have horn lengths in the range 38.774 to 40.426 cm (i.e. 39.600 cm ± 0.826). Furthermore, 95% of individuals should be within 2 standard deviations of the mean (i.e. 95% of topis should have a horn length between 37.948 and 41.252 cm).

Student's t test

Two sets of data can be compared using a t test. The calculation assesses whether there is a difference between the two data sets.

The formula for the t test is:

$$t = \frac{(\bar{x}_1 - \bar{x}_2)}{\sqrt{\left(\frac{\sigma_1^2}{n_1}\right) + \left(\frac{\sigma_2^2}{n_2}\right)}}$$

$\bar{x}_1, \bar{x}_2$ = mean of populations 1 and 2

σ_1, σ_2 = standard deviation of populations 1 and 2

n_1, n_2 = total number of values in samples 1 and 2

The null hypothesis and t value

The t test tells you whether a significant difference exists between two sets of data. The **null hypothesis** being tested is that no difference exists. Once you have calculated your t value, you need to:

1   Work out the **degrees of freedom** in the test (this is calculated as the total number of samples from both data sets, minus 2 $(n_1 + n_2 - 2)$).

2   Use a table of probabilities (which you will be given in your exam) to find a **probability value** for your *t* value, corresponding to the correct degrees of freedom.

3   The probability value tells you the likelihood of the difference between the two data sets being due to chance.

4   You can reject your null hypothesis if there is less than a 5% probability that the difference is due to chance. A probability (*p*) value of 0.05 or less tells you that there is a **significant difference** between the two groups you have been investigating.

Spearman's rank correlation coefficient

A correlation coefficient (r_s) is calculated to assess whether a relationship exists between two sets of rank-ordered data. The formula for calculating r_s is:

$$r_s = 1 - \frac{6\Sigma d^2}{n(n^2 - 1)}$$

where:

r_s = correlation coefficient

d = difference in ranks

Σ = the sum (total) of

n = number of pairs of data

The value of r_s will be between −1 and +1; a perfect negative correlation produces a value of −1, whereas a perfect positive correlation gives a value of +1. You will need to look up a probability (*p*) value for your r_s value (using the same procedure as you would with a *t* value) in order to conclude whether a correlation is significant. The critical values table for a Spearman's rank test is different from that of a *t*-test so you must check you are using the correct one.

> **Revision tip: Understand the formulae rather than memorising them**
>
> You are not expected to learn any of these formulae. However, you will be required to use them, if they are provided, and to understand what the results indicate.

Summary questions

1   State whether the following examples show discontinuous or continuous variation: **a** the volume of a protozoan cell **b** the presence or absence of haemophilia in a person **c** resting heart rate.   *(3 marks)*

2   Horn lengths were measured in two populations of topi.

	Number sampled	Mean horn length (cm)	Standard deviation (cm)
Population 1	10	41.200	0.930
Population 2	10	39.600	0.826

Use the *t* test to assess whether the two populations show a significant difference in horn length. For 18 degrees of freedom, a *t* value of 2.10 corresponds to a *p* value of 0.05.   *(5 marks)*

3   The mating success of 10 male topis was recorded along with their horn lengths.

Sample number	1	2	3	4	5	6	7	8	9	10
Horn length (cm)	38.400	39.000	39.200	40.100	39.400	38.900	40.200	39.200	40.800	40.800
Mean number of females mated per day	0.800	1.333	1.250	2.200	1.250	1.500	2.500	1.500	2.000	1.800

Calculate a Spearman's rank correlation coefficient to assess whether mating success is correlated with horn length in this population. For 18 degrees of freedom, an *r* value of 0.40 corresponds to a *p* value of 0.05.   *(6 marks)*

10.7 Adaptations

In the last two topics, you have explored variation between organisms. Any variation that benefits an organism in its environment is called an adaptation. Here you will learn about the types of adaptations exhibited by species. You will also look at situations in which unrelated species develop similar adaptations, which is known as convergent evolution.

Types of adaptation

Adaptations fall into three categories: anatomical, behavioural, and physiological.

Type of adaptation	Description	Roles of the adaptations: examples
Anatomical	Physical structures	**Communication** (e.g. species such as the flamboyant cuttlefish (*Metasepia pfefferi*) display bright colours to warn other species of their toxicity). **Locomotion** (e.g. fish species have evolved gas-filled swim bladders for buoyancy, fins, and streamlined shapes). **Feeding** (e.g. the structure of animals' teeth have evolved to suit their diets; herbivores have cusped molars for grinding whereas carnivores require sharp canines for tearing meat). **Water regulation** (e.g. some plant species, such as marram grass, need adaptations (thick waxy cuticles, curled leaves) to conserve water).
Behavioural	Simple, innate behaviours (such as reflexes) through to more complex (learned) behaviours	**Communication** (e.g. some animal species use elaborate courtship displays to attract mates). **Locomotion** (e.g. taxes, which are simple behaviours involving organisms moving in one direction, either towards or away from a stimulus; *E. coli* bacteria, for example, exhibit chemotaxis when they are attracted to certain chemicals). **Responding to seasonal changes** (e.g. migration of certain animal species and hibernation of mammalian species such as ground squirrels).
Physiological	Biochemical and cellular traits (e.g. the type of enzymes and hormones an organism produces)	**Feeding** (e.g. species have evolved different enzymes to digest the components of their diets). **Antibiotic resistance** (bacteria can evolve resistance to antibiotic drugs though a variety of molecular adaptations). **Adaptations to temperature** (e.g. icefish (*Chaenocephalus aceratus*) produce cold-resistant enzymes that are suited to their Antarctic climate).

Convergent evolution

Unrelated species can live in similar habitats and face similar selection pressures. This results in such species evolving, independent of each other, similar structures (known as analogous structures). This is called convergent evolution (because the phenotypes of unrelated species converge). Examples of convergent evolution include:

- Leaves, which have evolved in plants on several separate occasions.
- The ability to produce silk threads, which has evolved in weaver ants, spiders, and silk moths.
- Whales (order: Cetacea) and dugongs (order: Sirenia), which have evolved similar tail flukes.
- Echolocation (for hunting), which has evolved in bats, whales, and oilbirds.

Common misconception: Adaptation takes time

A species adapts over many generations. Adaptation is the process by which species evolve traits to suit their habitat. This tends to be a lengthy process, often taking millions of years. To speak of individual organisms *becoming* adapted is incorrect. Individuals *exhibit* adaptations (i.e. traits that have evolved over many generations).

Synoptic link

You learned about plant adaptations to prevent water loss in Topic 9.5, Plant adaptations to water availability.

Key term

Adaptation: A trait that benefits an organism in its environment and increases its chances of survival and reproduction.

Summary questions

1 State whether the following adaptations should be classified as anatomical, behavioural, or physiological: **a** sweating **b** phototaxis **c** opposable digits **d** lactose tolerance. *(4 marks)*

2 Aye-aye lemurs (Order: Primates) and striped possums (Order: Diprotodontia) both have elongated fingers that they use to locate invertebrates in trees. Explain why this is an example of convergent evolution. *(3 marks)*

3 Outline the anatomical, behavioural, and physiological adaptations of bacterial cells. *(6 marks)*

You have examined some of the evidence for evolution by natural selection in Topic 10.4, Evidence for evolution, and in the previous topic you learned about the types of adaptation that species can evolve. Here you will delve further into the mechanism of natural selection. You will also consider some recent examples of evolution and their implications for humans.

Natural selection

The characteristics of a population evolve over time through a mechanism called natural selection. The following steps are required:

1 **Variation** (due to genetic mutations) must exist within the population.

2 The presence of a **selection pressure** (a factor that affects an organism's chance of survival, such as the threat of predation, diseases, climate change, and changes in food availability).

3 Only individuals possessing traits enabling them to overcome the selection pressure will survive and reproduce (because they are better **adapted** than other organisms).

4 The gene variants (**alleles**) that enabled survival in the presence of the selection pressure are passed on to the next generation.

5 Over many generations the allele frequencies in the population will change and the balance of phenotypic characteristics will change accordingly to suit the population's habitat.

6 When a series of mutations arises in a population, natural selection can result in **speciation** (the formation of a new species).

Go further: A case of speedy evolution

Deer mice (*Peromyscus maniculatus*) are widespread across the USA. Most deer mice have a dark coat. However, in an area of Nebraska known as Sand Hills, a deer mouse population has evolved lighter fur to match the sandy soils in this habitat. The remarkable part of this story is that Sand Hills only formed 8000–15 000 years ago. The evolution of the deer mouse coat happened within the space of a few thousand years.

Researchers estimate that the allele for a sandy-coloured coat arose 4000 years ago. This produced variation in fur colour within the population. The deer mice would have faced a selection pressure in the form of predation (principally from birds). Mice with sandy fur would have been better camouflaged and less likely to be eaten. These individuals would have been better adapted than those with dark fur. Sandy mice would have been more likely to survive, reproduce, and pass on their alleles to offspring.

Over many generations, the frequency of the allele for sandy fur would have increased until the population comprised mainly sandy-coloured mice. The researchers calculated that the sandy coat allele gave these mice a 0.5% survival advantage.

1 Explain why this is an example of natural selection but not speciation.

2 Suggest why the allele frequencies for fur colour changed over time despite the sandy coat allele giving the mice only a 0.5% survival advantage.

Modern evolution: implications for humans

Some species evolve at a faster rate than others. The recent evolution of some species has implications for human populations. Some examples are outlined in the following table.

Species	Adaptation that has evolved	Implication for humans
Staphylococcus aureus	Antibiotic resistance (e.g. to methicillin)	Methicillin-resistant *Staphylococcus aureus* (MRSA) infections are very difficult to treat.
Flavobacterium sp	Production of nylonase enzyme	These bacteria can be used to remove factory waste.
Drosophila sp	Resistance to the insecticide malathion	*Drosophila* fruit flies can infest orange groves.

Summary questions

1 State three examples of a selection pressure that bacterial species may encounter. *(3 marks)*

2 Describe one example of an adaptation evolved in another species that has benefited humans. *(2 marks)*

3 Humans have developed chemicals called insecticides to kill undesired insect species. Explain how a species of insect could become resistant to an insecticide. *(4 marks)*

1 Which of the following is an example of a physiological adaptation in humans?

   A Tool use   **C** Lactose tolerance

   B Bipedalism   **D** Opposable digits   *(1 mark)*

2 Which of the following is a feature of prokaryotic organisms?

   A Membrane-bound organelles   **B** Chitin cell wall

   C 80S ribosomes   **D** Circular DNA   *(1 mark)*

3 For which of the following examples would the Student's *t* test be an appropriate statistical test to use?

   A Measuring the variation in trunk length in an African elephant (*Loxodonta africana*) population.

   B Comparing the mean fat reserves of two populations of grizzly bears (*Ursus arctos*).

   C Analysing the relationship between temperature and leaf length in common ivy (*Hedera helix*).   *(1 mark)*

4 The following passage concerns the process of natural selection. Complete the passage by choosing the most appropriate word to place in each gap.

   ……………. can produce new gene variants, known as …………….. . Some gene variants code for traits that provide a survival advantage when a population experiences a …………… ………………… (e.g. a new disease or a change in climate). Individuals that are best ……………… to their environment will have a higher probability of surviving and passing on their genes to the next generation.   *(4 marks)*

5 The graph below shows the range of heights in a sample of a human population.

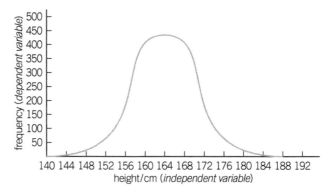

   a State the name given to this type of distribution.   *(1 mark)*

   b **i** What type of variation is shown in the graph?   *(1 mark)*

        ii Explain why height exhibits this type of variation.   *(2 marks)*

   c Estimate the mean, mode, and median height values from the graph.   *(2 marks)*

6 Describe the evidence that can be used to classify species.

   You should illustrate each type of evidence you include in your answer with examples.   *(6 marks)*

11.1 Biodiversity

Specification reference: 4.2.1(a)

Biodiversity, in a broad sense, is the variety of life in an area. We can, however, consider biodiversity on several scales: the genetic variation within one species, the variety of species in a habitat, or the range of habitats in an area.

Measures of biodiversity

	Genetic diversity	Species diversity	Habitat/ecosystem diversity
What is measured?	An estimate of gene variants (alleles) in a species.	Two components are measured: **Species richness** (the number of species in an area). **Species evenness** (the number of individuals of each species).	The number of different habitats (or ecosystems) in an area. This is the hardest measure of biodiversity to calculate because the boundaries of ecosystems are often difficult to determine with accuracy.
Topic reference	11.5, Calculating genetic biodiversity	11.4, Calculating biodiversity	11.6, Factors affecting biodiversity

Common misconception: Ecological terms

In this chapter you will be wading through several related ecological terms, which are easy to confuse. For example, a **population** is a group of organisms of the *same* species. A **community** is a collection of populations in a **habitat**, which is the natural environment of organisms. An **ecosystem** is formed from a community of species (the **biotic** component) and their non-living surroundings (the **abiotic** components such as air, water, and soil) interacting together.

Revision tip: Richness and evenness

An ecosystem is considered biodiverse only when species richness and evenness are both high. For example, many species may be identified in a habitat, indicating species richness. Yet if one or two of the species dominate then species evenness is low.

Summary questions

1 Describe the difference between a population and a community. *(2 marks)*

2 Suggest and explain which measure of biodiversity is most significant when assessing the health of one ecosystem. *(2 marks)*

3 Explain how a habitat can contain many species but its biodiversity is considered low. *(3 marks)*

Ecologists can monitor biodiversity by sampling parts of an ecosystem. Counting every single organism is impossible, but calculating the biodiversity in small sections of an ecosystem enables the overall biodiversity to be estimated.

The principles of sampling

The choice of sampling method is an important decision. You should bear in mind a few key principles when designing your sampling method.

- **Larger sample sizes** are more representative of the whole ecosystem.
- **Avoiding bias** when choosing where to take samples will increase the **validity** of your results.
- Sampling, no matter how good, can only be claimed to represent a close **estimate** of an ecosystem's biodiversity.

Random and non-random sampling

The design of your sampling method will depend on the ecosystem being studied. Two approaches to assessing biodiversity are **random sampling** and **non-random sampling**.

Random sampling

You can decide the location of sampling points in an ecosystem by:

- generating random numbers, which are used as grid coordinates
- taking samples from these coordinates.

Random sampling avoids bias, but it can produce an unrepresentative picture of an ecosystem. This is especially true if the surveyed area is large. In addition, some species may be unevenly distributed and found only in certain parts of the ecosystem. Random sampling could miss these species, especially if your sample size is small because of time constraints. **Stratified sampling** can overcome this problem.

Non-random sampling

Stratified sampling

The study site is divided into smaller areas, based on the distribution of habitats. This method ensures species are not overlooked. The sampling is **more representative** of the ecosystem and **reduces sampling error**.

 Worked example: Stratified sampling

Dense tree growth represents 80% of a woodland ecosystem and the remaining 20% consists of a shrubby clearing. These two areas are likely to contain different communities. If you plan to take 20 samples, 16 should be from the area with trees and four should be from the clearing. The number of samples is therefore proportional to the size of each area. Within each sub-region of your ecosystem, the location of each sample is decided randomly.

Systematic sampling

A **transect** is used where environmental gradients exist (e.g. soil pH or light intensity changes across a habitat). You can investigate whether the distribution of organisms also changes across the habitat.

11.3 Sampling techniques

Specification reference: 4.2.1(b)(ii) and (c)

Once a sampling strategy has been chosen, the techniques and equipment used for sampling will depend on the type of organisms being sampled.

Techniques for sampling organisms

▼ **Table 1** *Techniques and equipment for sampling animals*

Technique	Which animals are sampled?	How is it used?
Pooter	Insects	Insects are sucked into a chamber
Sweep nets	Insects in long grass	The net is swept through the habitat
Pitfall traps	Small, crawling invertebrates	A hole in the ground traps organisms
Tree beating	Tree-dwelling invertebrates	Trees are shaken and organisms fall on a sheet
Kick sampling	River-dwelling organisms	A river bed is disturbed and organisms are captured in a net

Plants are sampled using **quadrats**, which are square frames. The species present in the quadrat can be observed, identified with a key, and counted.

▼ **Table 2** *The two quadrat designs for sampling plants*

Quadrat design	How is it used?
Point quadrat	Pins are pushed through holes in a bar that spans the quadrat. All species that touch the pins are identified and recorded.
Frame quadrat	The quadrat is divided into a grid of smaller squares. Species are identified and their abundance can be estimated.

Measuring abiotic factors

Ecologists often measure abiotic (environmental) factors in the areas where they conduct species sampling.

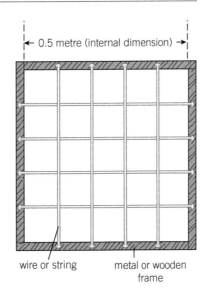

← 0.5 metre (internal dimension) →

wire or string   metal or wooden frame

▲ **Figure 1** *A frame quadrat*

▼ **Table 3** *Examples of abiotic factors that can be measured*

Abiotic factor	Sensor	Units of measurement
Wind speed	Anemometer	$m\,s^{-1}$
Light intensity	Photometer	lux
Temperature	Thermometer	°C

Revision tip: Richness and evenness

An accurate recording of species **richness** (an area's total number of species) requires a good **identification key**. Species **evenness** is a comparison of species' numbers; this requires an accurate method for **measuring the abundance** of each species.

Quadrats can be used to record species abundance in two ways:

1 Count the *absolute* number of individuals in the quadrat (i.e. density per m^2 is recorded).

2 *Estimate* the percentage of a quadrat covered by a species.

Revision tip: Drop the 'n'

Ecological sampling is carried out with a quadrat. Remember to leave out the 'n'!

Summary questions

1 Pitfall traps are covered. Suggest why. (*1 mark*)

2 Suggest why the percentage cover of a species rather than its density might be recorded. (*2 marks*)

3 An anemometer measured a wind speed of $6\,m\,s^{-1}$. Express this speed in $km\,hr^{-1}$. Give your answer to 2 significant figures. (*2 marks*)

11.4 Calculating biodiversity

Specification reference: 4.2.1(d)

Summary questions

1 State the difference between n and N in the Simpson's Diversity Index formula. (*2 marks*)

2 Explain why an ecosystem with a low value of D is vulnerable to environmental change. (*3 marks*)

3 Calculate the Simpson's Diversity Index for the two habitats in the following table. Which habitat is more likely to be sensitive to environmental change? (*3 marks*)

Species	Habitat A	Habitat B
Woodrush	1	4
Holly (seedlings)	11	7
Bramble	1	3
Yorkshire Fog	1	2
Sedge	2	3

Species diversity is a measure of the number of species in a habitat (species richness) and the relative abundance of individuals in each of these species (species evenness). Both factors must be taken into account in the calculation of species diversity.

Calculating species biodiversity

Species diversity is calculated using the Simpson's Diversity Index.

Simpson's Diversity Index $(D) = 1 - \left[\Sigma \left(\frac{n}{N} \right)^2 \right]$

where n is the number of individuals of a particular species and N is the total number of all individuals of all species. Remember, Σ is the symbol for 'sum'. In this case, we need to calculate $\left(\frac{n}{N} \right)^2$ for each species and then add the values together.

Worked example: species diversity calculation

	Field A (adjacent to intensively farmed land)			Field B (adjacent to extensively farmed land)		
	n	$\frac{n}{N}$	$\left(\frac{n}{N}\right)^2$	n	$\frac{n}{N}$	$\left(\frac{n}{N}\right)^2$
Cocksfoot grass	60	0.60	0.3600	40	0.40	0.1600
Timothy grass	29	0.29	0.0841	20	0.20	0.0400
Meadow buttercup	4	0.04	0.0016	13	0.13	0.0169
Cowslip	1	0.01	0.0001	8	0.08	0.0064
White clover	2	0.02	0.0004	13	0.13	0.0169
Dandelion	4	0.04	0.0016	6	0.06	0.0036
Σ	–	–	0.4478	–	–	0.2438
$1-\Sigma$	–	–	0.5522	–	–	0.7562

In this example, n is the percentage cover of each plant species. N is therefore 100.

A high biodiversity is reflected by a high value of Simpson's index. Low values indicate an ecosystem is dominated by a few species and is unstable. In this example, Field B has the higher Simpson's index value and we can conclude that its biodiversity is greater than that of Field A.

11.5 Calculating genetic biodiversity

Specification reference: 4.2.1(e)

Genetic diversity determines how easily a species is able to adapt to changes in its environment. As you learned in Topic 10.8, Changing population characteristics, variation is necessary for evolution to occur. A species with high genetic diversity will contain a wide range of traits on which natural selection can act. Therefore genetic diversity increases the chance of a species adapting and surviving to environmental changes. There are several ways of measuring the genetic diversity of a species, which we discuss here.

Factors affecting genetic diversity

The genetic diversity of a population can increase (i.e. the number of alleles in the gene pool increases) due to:

- DNA mutation (see Topic 10.5, Types of variation)
- Gene flow from another population (i.e. breeding between populations of the same species).

Genetic diversity decreases due to:

- Selective breeding
- Captive breeding
- Genetic bottlenecks (i.e. when a population is reduced to a small size because of disease, habitat destruction, or migration).

Measuring genetic diversity

You read about the genetic code in Topic 3.9, DNA replication and the genetic code. A gene is a DNA sequence that codes for a polypeptide. Members of a species have the same genes, but they can have different versions of the same gene. These gene variants are known as **alleles**.

Genetic diversity is determined by the variation in a species' genes. A species that has a high percentage of genes with only one possible variant has low genetic diversity. A species that has a high percentage of genes with several possible variants has high genetic diversity.

Several methods can be used to calculate genetic diversity:

- the number of alleles per gene
- heterozygosity: the proportion of individuals in a population that have two different alleles for a particular gene
- the proportion of genes for which more than one allele exists. A gene that has two or more possible variants/alleles is known as a **polymorphic gene**. A gene for which only one variant/allele exists is called a monomorphic gene and the calculation is:

$$\text{Proportion of polymorphic genes} = \frac{\text{Number of polymorphic genes}}{\text{Total number of genes}}$$

Key term

Alleles: Different versions of the same gene (i.e. gene variants).

Revision tip: 'Total' means 'total sampled'

You should note that the 'total number of genes' used in the calculation tends to be a small sample of genes rather than every gene in a species' genome.

⊞ Worked example: Genetic diversity calculation

The following table shows genetic data for three populations of monkeys.

	Blue monkey population	Vervet monkey population 1	Vervet monkey population 2
Sample size (number of individuals)	93	124	364
Number of genes studied	33	23	18
Number of polymorphic genes	5	4	3
Average heterozygosity (%)	4.6	5.6	4.0

Q1 Is it possible to conclude which population has the highest genetic diversity?

Q2 What could be done to improve our confidence in the results?

A1 The proportions of polymorphic genes in each population are:

Blue monkeys: $\dfrac{5}{33} = 0.15$ (or 15%)

Vervet monkey population 1: $\dfrac{4}{23} = 0.17$ (or 17%)

Vervet monkey population 2: $\dfrac{3}{18} = 0.17$ (or 17%)

Based on average heterozygosity, population 1 of vervet monkeys has the highest genetic diversity. However, based on the proportion of polymorphic genes, the two vervet monkey populations have the same genetic diversity.

A2 Increasing the sample sizes and analysing a greater number of genes would improve confidence levels in the results. The genes studied represent approximately 0.1% of the total number of genes in these species. Studying more genes will increase the accuracy of the genetic diversity estimates.

Summary questions

1 Describe how genetic diversity can decrease in a natural population. *(3 marks)*

2 Explain the importance of genetic diversity to species survival. *(2 marks)*

3 Complete the following table and discuss which species has the greatest genetic diversity. *(5 marks)*

Species	Number of monomorphic genes studied	Number of polymorphic genes studied	Percentage of polymorphic genes (%)	Average heterozygosity (%)
A	28	8		4.2
B		10	22	7.0
C	8	3		5.4

11.6 Factors affecting biodiversity

Specification reference: 4.2.1(f)

Our impact on ecosystems is a more significant issue now than at any other time in history because of the continuing rise in the global human population. In this topic you will examine the direct and indirect impacts of human activities on biodiversity.

Human impacts on biodiversity

Human activity	Direct impacts on biodiversity	Indirect impacts on biodiversity
Forestry management and deforestation	Removal of native trees (e.g. deciduous woodland in the UK) has **destroyed habitats**, causing animal migrations or death. Sometimes non-native trees are grown instead to meet the demands of the timber and fuel industries, but biodiversity is lowered because fewer species are grown.	
Agriculture	**Deforestation** to clear land for farming. **Hedgerow removal** (to maximise space available for crops and machinery). **Intensive farming** uses chemicals (e.g. **pesticides and herbicides**) to kill plants and animals that are considered pests or weeds. For example, **neonicotinoid insecticides** have been blamed for reducing populations of bee species. **Monocultures** (growing a single crop) reduce the number of animals that can use the farmland as a habitat.	Fertilisers can pass into watercourses, contaminating neighbouring ecosystems (e.g. **eutrophication**). The demand for water (e.g. for irrigation of farmland) has resulted in rivers and lakes being **drained**, decimating aquatic ecosystems. For example, the Aral Sea has shrunk to 10% of its original volume due to irrigation schemes.
Fossil fuel combustion	Habitat destruction for access to fossil fuel sources	**Acid rain** production (due to air pollution) has contributed to habitat destruction (e.g. deforestation). **Climate change** (due to greenhouse gas emissions) has caused habitat destruction (e.g. desertification, melting ice in Arctic habitats, and changes in sea temperatures and currents).

Revision tip: Eutrophication

Eutrophication results from an increase in nutrients in freshwater lakes and rivers. These nutrients can enter the water from sewage or fertilisers. The nutrients encourage the growth of algae, which prevents sunlight reaching deeper aquatic plants. When these plants die, decomposers deplete oxygen in the water, which further decreases biodiversity.

Summary questions

1. State two ways in which climate change has damaged ecosystems. *(2 marks)*

2. Describe how the rising human population has affected biodiversity in aquatic ecosystems. *(4 marks)*

3. Intensive farming is characterised by the use of chemicals (e.g. fertilisers and pesticides), high costs and a large input of labour. This contrasts with extensive farming, which requires fewer chemicals, and less money and labour. Contrast the advantages and disadvantages of the two forms of farming. *(3 marks)*

11.7 Reasons for maintaining biodiversity

Specification reference: 4.2.1 (g)

From a human perspective, the maintenance of biodiversity is important for a number of economic, ecological, and aesthetic reasons, which we examine here.

Why is biodiversity important?

Aesthetic reasons

Plant and animal life provides enrichment, relaxation, and inspiration; we all enjoy looking at a panda cub or strolling through beautiful woodland.

Economic reasons

- **Discovery of useful genes or compounds:** approximately half of our medicines contain a chemical derived from animal or plant species. Extinct species may have possessed genes that would have been useful in medical research or agriculture in the future.
- **Sustainable** removal of resources helps land remain economically viable for longer.
- High genetic diversity increases the chance of species (e.g. crop species) **adapting to future environmental change**. Greater biodiversity increases the range of traits available for **artificial selection** in the future.
- Money can be made through **tourism** in biodiverse areas.

Ecological reasons

As you learned in Topic 11.4, Calculating biodiversity, ecosystems with low biodiversity are unstable and especially vulnerable to environmental change. Species within a community are interdependent; the removal of one species can disrupt the rest of its food chain. For example, the extinction of a food source is likely to reduce the populations of its predators. The disappearance of insect pollinators (e.g. bee species) is liable to have a negative impact on populations of flowering plants.

> **Revision tip: Sustainability**
>
> Sustainability is the concept of human populations meeting energy and food requirements without compromising biodiversity (or the ability to meet requirements in the future).

> **Revision tip: Keystone species**
>
> A keystone is the stone that allows an arch to be self-supporting and prevents it collapsing. A keystone species prevents an ecosystem from collapsing—it has a significant effect on other species. An ecosystem is dramatically altered if a keystone species is removed.
>
> For example, for part of the year, acorn banksia trees are the only source of nectar for honeyeater birds. Later in the year, honeyeaters pollinate other plants. The acorn banksia is a keystone species because the honeyeater population (and therefore many plant populations) would dwindle without it.

> **Summary questions**
>
> 1 Outline two economic reasons for maintaining biodiversity. *(2 marks)*
>
> 2 Suggest two long-term benefits of sustainable forestry. *(2 marks)*
>
> 3 Sea otters eat urchins, which eat kelp (a species of seaweed). Explain why sea otters are a keystone species in kelp forest habitats. *(2 marks)*

11.8 Methods of maintaining biodiversity

Specification reference: 4.2.1(h) and (i)

Human populations can stem the rate of species extinctions by reducing levels of deforestation and intensive agriculture, and limiting the rate of global warming. However, more proactive approaches can be used to maintain or restore biodiversity. These conservation methods are our focus in this topic.

Methods for maintaining biodiversity

Scientists can either conserve species *in situ* (i.e. in their natural habitat) or *ex situ* (i.e. outside their natural habitat).

▼ **Table 1** *In situ conservation methods*

In situ method	Possible features
Wildlife reserves	Human access is restricted (e.g. poaching is banned or regulated)
	Animals are fed
	Reintroducing species
	Culling (i.e. removing invasive species)
	Preventing succession, which is a natural process of ecological change that occurs over many years (e.g. the use of controlled grazing maintains heathland ecosystems and prevents the process of succession, which would result in the formation of woodland)
Marine conservation zones	Limits placed on hunting and fishing within refuge areas

▼ **Table 2** *Ex situ conservation methods*

Ex situ method	Possible features
Botanic gardens	Plants experience optimum conditions (e.g. ideal soil nutrients and a lack of pests)
Seed banks	Seeds from many plant species are stored in conditions that enable them to remain viable for centuries
Captive breeding programmes (e.g. in zoos or aquatic centres)	Production of offspring in human-controlled environments
	Reintroduction of species into natural habitats
	Animals experience optimum nutrition, a lack of predators, and medical attention

> **Revision tip: Conserve or preserve?**
> 'Conservation' and 'preservation' are two terms with subtle differences. Conservation is the active and sustainable management of an ecosystem, whereas preservation leaves an ecosystem undisturbed (i.e. without human intervention).

Conservation agreements

National and international agreements enable conservation targets and rules to be created. Important agreements include:

- The Convention on International Trade in Endangered Species (**CITES**) – regulates the trade of more than 35 000 species.

- The **Rio Convention** – targets were agreed for sustainability, reducing desertification, and limiting greenhouse gas emissions.

Summary questions

1 State the principal outcomes of the Rio Convention. *(2 marks)*

2 Suggest three advantages of using seed banks rather than botanic gardens to conserve plant species. *(3 marks)*

3 Evaluate the advantages and disadvantages of *ex situ* conservation in comparison to *in situ* methods. *(5 marks)*

1 The genetic diversity of four populations of a species was being studied. Data for the four populations are shown below.

Population	Number of loci studied	Number of monomorphic genes
A	45	37
B	42	38
C	20	16
D	33	22

a Based on the data in the table, which population has the highest genetic diversity? Show your working and explain your conclusion.

(3 marks)

b Comment on the reliability of conclusions drawn from this data.

(2 marks)

2 More than 95% of the species that have existed on Earth during the past 3.5 billion years are now extinct.

a In 2014 scientists calculated the relative threat posed to biodiversity by different human activities. 37% of the total threat was estimated to be from human exploitation of ecosystems (e.g. fishing and hunting).

State two other threats to biodiversity posed by human activity.

(2 marks)

b The highest estimate for the extinction rate of species is 0.7% of existing species per year.

Assuming an extinction rate of 0.7% per year and a current global species count of 5 million species, how many species will exist after
i 1 year   *(1 mark)*   ii 3 years   *(1 mark)*

Give your answers to 4 significant figures.

c The International Union for Conservation of Nature (IUCN) decide which species are considered to be threatened with extinction. For example, 26% of mammals were placed on the IUCN's Red List of Threatened Species in 2014.

Describe how sampling techniques and data analysis can be used to monitor the population size of a mammalian species to decide whether it should be placed on the Red List of Threatened Species.   *(6 marks)*

3 Which of the following statements best describes stratified sampling?

A Selecting sampling locations at random.

B Sampling based on availability.

C Sampling sub-groups in proportion to their relative sizes.

D The use of a transect for sampling.   *(1 mark)*

4 Which of the following sampling techniques would be most appropriate to use when sampling river-dwelling invertebrates?

A Kick sampling   C Quadrats

B Pooters   D Pitfall traps   *(1 mark)*

5 Which of the following is a property of a habitat with a low value of Simpson's Diversity Index?

A A large number of species

B Few ecological niches

C Complex food webs

D Environmental change produces small effects   *(1 mark)*

12.1 Animal and plant pathogens

Specification reference: 4.1.1(a)

This chapter focuses on communicable diseases, which are caused by infective microorganisms known as **pathogens**.

Types of pathogen

Pathogens can be categorised into four groups: bacteria, fungi, protoctista, and viruses.

Type of pathogen	Mode of action	Appearance	Examples of diseases
Bacteria	Disease symptoms are often caused by **toxin production**	**Prokaryotic** cells (see Topic 2.6, Prokaryotic and eukaryotic cells) Shapes include rod (bacilli), spherical (cocci) and spiral	Tuberculosis (TB) Bacterial meningitis Ring rot
Fungi	They **secrete enzymes** that digest living cells, enabling the fungus to spread through tissue	**Eukaryotic** organisms (see Topic 2.4, Eukaryotic cell structure)	Ring worm Black sigatoka
Protoctista	They often consume the cell material of their host	**Eukaryotic** cells	Malaria Potato blight
Viruses	They **insert genetic material** into their host's DNA, taking control of cell metabolism	Usually considered **non-living** Protein coat enclosing genetic material	Influenza Tobacco mosaic virus

Revision tip: Are you alive?

Viruses lack many traits that define living organisms (e.g. they cannot grow, synthesise proteins, or reproduce independently, and they lack membranes). They lead a 'borrowed life', relying on host cells to reproduce, and exploiting their metabolism. Some scientists think viruses occupy a grey area between living and non-living.

Revision tip: We're friendly, really ...

Most species of bacteria, protoctista, and fungi do not cause disease—only a small number of species are pathogenic.

Revision tip: Please, use my full title ...

Try to use the full binomial name of pathogens (e.g. *Mycobacterium tuberculosis*). Abbreviations are acceptable once the full name has been used once.

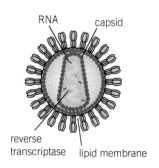

▲ **Figure 1** *The structure of HIV (Human Immunodeficiency Virus)*

Summary questions

1 State two eukaryotic kingdoms that contain pathogenic species. *(2 marks)*

2 Describe the typical cause of symptoms for diseases resulting from
  a bacterial infection **b** fungal infection. *(2 marks)*

3 Evaluate whether viruses should be considered organisms. *(3 marks)*

12.2 Animal and plant diseases

Specification reference: 4.1.1(a)

You were introduced to the four pathogenic taxa in Topic 12.1, Animal and plant pathogens. Now you will learn about some of the diseases that can be caused by these pathogens.

Plant diseases

▼ **Table 1** *Diseases of plant species*

Disease	Pathogen	Symptoms
Potato blight	*Phytophthora infestans* (a protoctist)	Hyphae (branching structures) penetrate cells, destroying tubers, leaves, and fruit
Ring rot	*Clavibacter michiganensis* (a bacterium)	Destroys vascular tissue in leaves and tubers
Tobacco mosaic virus	TMV (a virus)	Mosaic patterns of discoloration on leaves, flowers, and fruit
Black sigatoka	*Mycosphaerella fijiensis* (a fungus)	Hyphae penetrate and digest leaf cells, turning leaves black

Animal diseases

▼ **Table 2** *Diseases of animal species*

Disease	Pathogen	Symptoms
Malaria	*Plasmodium spp.* (protoctists)	Infects erythrocytes and liver cells, causing fever and fatigue
Tuberculosis (TB)	*Mycobacterium tuberculosis* (a bacterium)	Destroys lung tissue, resulting in coughing, fatigue, and chest pain
HIV/AIDS	Human immunodeficiency virus (HIV)	Infects T helper cells (see Topic 12.6, The specific immune system), thereby inhibiting the immune system
Athlete's foot	*Tinia pedia* (a fungus)	Digests skin on people's feet, causing cracking and itchiness

Revision tip: Retroviruses

HIV is a retrovirus, which means it contains RNA rather than DNA. It also contains an enzyme called reverse transcriptase, which produces a DNA copy of its RNA genome. The viral DNA is incorporated into the DNA of a T helper cell. Copies of the virus can then be produced.

Go further: Ebola

Ebola is a disease caused by a virus. It causes internal bleeding and kills approximately 50% of those infected, on average. The disease was first described in human populations in 1976. More than 20 000 cases were diagnosed in 2014, which represents an epidemic. The virus is communicable, but can be passed between humans only through direct contact with body fluids.

Suggest the difficulties that will be encountered when treating diseases such as Ebola, which is caused by a pathogen that has only recently evolved to infect humans.

Summary questions

1 State one similarity and one difference between potato blight and black sigatoka. *(3 marks)*

2 Describe how HIV is able to replicate. *(4 marks)*

3 A new strain of the H1N1 virus caused a pandemic (a worldwide outbreak) of influenza in 2009. Suggest why the new strain resulted in a pandemic. *(2 marks)*

12.3 The transmission of communicable diseases

Specification reference: 4.1.1(b)

You have learned that a communicable disease is one caused by pathogens, which can be transmitted between organisms. The subject of this topic is *how* these pathogens are passed on. Various modes of transmission exist, which can be categorised as direct or indirect.

Pathogen transmission between animals

▼ **Table 1** *Animal pathogens: modes of transmission*

Mode of transmission		Description	Examples
Direct	Contact	Contact with skin, or body fluids	Bacterial meningitis
	Entry through the skin	E.g. wounds, bites, or infected needles	HIV/AIDS Septicaemia
	Ingestion	Consumption of contaminated food or drink	Amoebic dysentery
Indirect	Fomites	Inanimate objects (e.g. bedding or clothes) that transfer pathogens	Athlete's foot
	Inhalation	Breathing in droplets containing pathogens	Influenza
	Vectors	Anything that carries a pathogen from one host to another is a vector (e.g. water, and many different animals)	Malaria (vector = mosquitoes)

Pathogen transmission between plants

▼ **Table 2** *Plant pathogens: modes of transmission*

Mode of transmission		Description	Examples
Direct	Contact	Contact between a healthy plant and a diseased plant	TMV Potato blight
Indirect	Soil contamination	Pathogens, or reproductive spores, move into the soil from infected plants	Black sigatoka Ring rot
	Vectors	Wind, water, and animals can act as vectors to transmit plant pathogens	*P. infestans* spores can be carried by air currents, causing blight to spread

Summary questions

1   Outline the social and economic factors that increase the risk of a communicable disease being spread. *(4 marks)*

2   Explain what is meant by a fomite, and state two diseases that are contracted through fomite contact. *(3 marks)*

3   Suggest why potatoes cannot be grown for at least two years on land that has supported plants with ring rot. *(2 marks)*

12.4 Plant defences against pathogens

Plants, as you have seen already in this chapter, can be vulnerable to infections by pathogens. In response, plants have evolved a range of defences to fend off their pathogenic attackers.

Types of defence

Physical defence

Callose is a polysaccharide formed from β-glucose monomers, joined with 1,3 glycosidic bonds (and some 1,6 linkages). It is largely linear (with a few branches) but helical. Callose is produced in response to pathogenic attacks and deposited in cell walls, plasmodesmata (i.e. pores in cell walls), and in sieve plates. The callose acts as a barrier to prevent further infection.

Chemical defence

Plants have evolved a range of chemical defences against pathogens and pests.

▼ **Table 1** *Plant chemical defences against pests and pathogens*

Type of chemical defence	Examples
Insect repellents	Citronella, produced by lemon grass
Insecticides	Pyrethrins, produced by chrysanthemums
Antibacterial compounds	Gossypol, produced by cotton
Antifungal compounds	Saponins, produced by many species (e.g. soapworts)
Anti-oomycetes	Glucanase enzymes, which destroy cell walls in *P. infestans*
General toxins	Cyanide compounds

> **Revision tip: Recognition and response**
> Receptors in plant cells recognise molecules on (or produced by) pathogens. This triggers a cascade of reactions, which switches on genes to produce defensive chemicals and molecules such as callose.

Summary questions

1 State two types of chemical defence produced by plants against pathogens. *(2 marks)*

2 Suggest why callose is deposited in **a** cell walls **b** plasmodesmata **c** sieve plates during an attack by a pathogen. *(3 marks)*

3 Compare the structures of callose and cellulose. *(4 marks)*

12.5 Non-specific animal defences against pathogens

Specification reference: 4.1.1(d) and (e)

Like plants, animals have evolved a range of defences against pathogens. These defences can be divided into mechanisms to combat specific pathogens and non-specific defences, which have evolved to repel a wide range of pathogens. This topic focuses on non-specific defences against pathogens.

Primary defences

Primary defences are the barriers that prevent pathogens from entering the body. They include: the **skin**, the **conjunctiva** (membrane covering the eye), **mucus**, and **ciliated epithelia** in airways, and the mucus layer and **acidic conditions** in the stomach and vagina.

Repairing the primary defences: blood clotting

Cuts to the skin leave an organism open to infection. The evolution of a blood clotting system enables repairs to be made to primary defences whenever they are damaged.

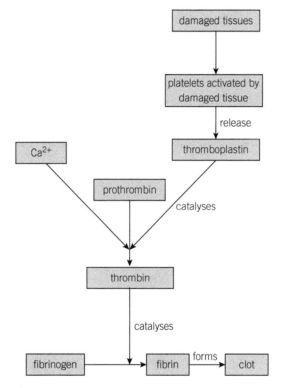

▲ **Figure 1** *The cascade of reactions that lead to blood clotting*

Secondary defences: inflammation and phagocytes

Pathogens can evade primary defences and infect an animal. Infections trigger internal non-specific responses: inflammation and phagocytosis.

▼ **Table 1** *The key features of inflammation and phagocytosis*

	What is the process?	How does it help?
Inflammation	**Mast cells** (which are leucocytes) release **histamines**, which dilate blood vessels and cause more plasma to move into tissue fluid. This **raises temperature** and causes **swelling**.	High temperature reduces the rate of pathogen reproduction. Inflammation is thought to be protective (e.g. isolating pathogens).
Phagocytosis	The phagocyte engulfs the pathogen The pathogen is enclosed in a vacuole (called a **phagosome**) **Lysosome** fuses with phagosome (forms a **phagolysosome**) **Enzymes** released by the lysosome digest the pathogen	Destruction of pathogenic cells

Synoptic link

You were introduced to the concept of phagocytosis in Topic 5.4, Active transport.

Revision tip: Helpful chemicals

Two sets of molecules assist phagocytes: cytokines and opsonins. **Cytokines** are cell-signalling molecules that, among other roles, attract phagocytes to sites of infection. **Opsonins** bind to pathogens and mark them for phagocytosis. Phagocytes have receptors that bind to opsonins.

Summary questions

1   Why are primary defences and phagocytosis known as non-specific defences?   (*1 mark*)

2   Explain the roles of thromboplastin and thrombin in blood clotting.   (*2 marks*)

3   Explain the role played by mast cells in defence against pathogens.   (*3 marks*)

In animals, non-specific defences, such as phagocytosis, combat pathogens as soon as they infect an organism. The specific immune system targets particular pathogens but takes longer to respond.

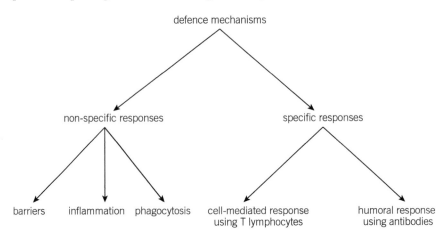

▲ **Figure 1** *An overview of animal defence mechanisms against pathogens*

Lymphocytes

Lymphocytes are white blood cells (leucocytes) that perform a variety of roles within the specific immune system.

▼ **Table 1** *Types of T and B lymphocyte*

Type of lymphocyte		Role
T lymphocyte	T helper cells	Produce **cytokines**, which stimulate B cells and other T cells
	T killer cells	Produce **perforin**, which damages the cell membranes of pathogens
	T memory cells	Recognise antigens from previous infections (**immunological memory**)
	T regulator cells	Control the immune system (**preventing autoimmune responses**)
B lymphocyte	Plasma cells	Produce **antibodies**
	B effector cells	Divide to **form plasma cell** clones
	B memory cells	Remember specific antigen (enables rapid **secondary immune response**)

Cell-mediated vs humoral immune responses

Specific immune responses are either cell-mediated or humoral.

▼ **Table 2** *The two branches of the specific immune system*

	Cell-mediated immunity	Humoral immunity
What happens?	**Antigen-presenting cells** (e.g. phagocytes) activate **T-helper cells**, which stimulate **phagocytosis**, and T memory and **killer cell** production No antibodies	**Clonal selection** of antigen-specific B cell **Clonal expansion** to produce **plasma cells** and B memory cells **Antibody production**
Typical targets	Viruses and cancerous cells	Bacteria and fungi

Revision tip: Good sense of humor?

The humoral immune response does not involve stand-up comedy; 'humor' is an old-fashioned word for body fluid, and the humoral response takes place in blood plasma and tissue fluid.

Antibodies

Antigens are molecules on cells (or viruses) that the immune system can use to detect infection. An organism's own cells have *self* antigens, whereas pathogens exhibit *non-self* antigens. **Plasma cells** in the immune system produce **antibodies** that are specific to a pathogen's non-self antigens.

The variable region is different in each antibody, which enables each antibody to bind to a specific antigen (on a pathogen). Depending on the type of antibody, a variety of fates await the pathogen.

▼ **Table 3** *Types of antibody defence*

Method of antibody defence	What happens?
Opsonisation	Antibody acts as an opsonin (speeding up phagocytosis).
Agglutination	Antigen–antibody complexes clump together. This clump is too large to enter cells and enables phagocytes to engulf several pathogens at once.
Neutralisation	Antibodies bind to toxins, rendering them harmless.

Autoimmune diseases

The immune system can malfunction and stop recognising self antigens. The body's cells are attacked by its own immune system. This is known as **autoimmune disease**.

▼ **Table 4** *Examples of autoimmune diseases*

Autoimmune disease	Body part affected	Symptoms
Grave's disease	Thyroid gland	Overactive thyroid, causing weight loss and muscle weakness
Vitiligo	Melanocytes	Loss of skin pigmentation
Type 1 diabetes	Pancreatic β-cells	Lack of insulin production; loss of blood glucose regulation

Summary questions

1 Outline the role of T killer cells. *(2 marks)*

2 Explain how agglutination limits bacterial infection of cells. *(3 marks)*

3 Suggest why someone with HIV/AIDS may not exhibit a secondary immune response despite already encountering a pathogen. *(2 marks)*

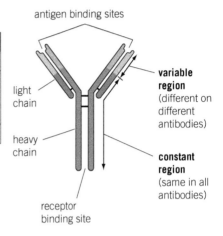

▲ **Figure 2** *The structure of an antibody (a glycoprotein comprising four polypeptide chains)*

12.7 Preventing and treating disease

Specification reference: 4.1.1(j), (l), (m), and (n)

An animal is immune if they can be infected with a pathogen without developing any symptoms of the disease. This immunity can develop in a number of ways, which you will read about here. This topic also examines methods of treating diseases once they have been contracted (e.g. the use of antibiotics).

Vaccinations

The principle of vaccination is to persuade the body to produce antibodies and memory cells against a particular pathogen without a person contracting the disease. Vaccination of many people in a population can prevent a disease spreading; this is called **herd immunity** and prevents **epidemics**.

▼ **Table 1** *The different ways in which a person can gain immunity*

	Natural	Artificial
Active	Memory cells produced following pathogenic **infection**	Memory cells produced following a **vaccination** (see below)
Passive	Fetal immunity (**maternal antibodies** cross the placenta)	**Antibodies are injected** into a person, providing temporary immunity

▼ **Table 2** *Types of vaccine*

Type of vaccine	How does it work?	Examples of diseases	Advantages/ disadvantages
Weakened, live pathogen	Modified pathogen that is alive but not pathogenic	Mumps, polio, measles, TB	Strongest response and long-lasting immunity, but (rarely) organism may revert and become pathogenic
Dead/inactivated pathogen	Pathogen is killed but its antigens are still present	Influenza, whooping cough	Stable and safer than live vaccines, but response is weaker (boosters required)
Toxoids	Modified toxins	Tetanus, diphtheria	Safe, but may not give strong response
Subunits	Isolated antigens	HIB	Vaccines for several strains produced

Sources of medicines

Vaccines are designed to prevent pathogenic infections. When someone does contract a disease, however, medicinal treatments can be given. Medicinal drugs are often derived from natural compounds.

▼ **Table 3** *Medicines from natural sources*

Medicine	Source	Properties and uses
Quinine	*Cinchona* spp.	Antimalarial, painkilling
Aspirin	*Salix alba* (willow)	Anti-inflammatory, painkilling
Penicillin	*Penicillium* fungi	Antibiotic

Antibiotic resistance

Mutation can result in the evolution of bacteria that are resistant to antibiotics (e.g. methicillin-resistant *Staphylococcus aureus* (MRSA)). The spread of antibiotic-resistant infection can be reduced by minimising the use of antibiotics (as overuse can accelerate natural selection of resistant strains) and using good hygiene practices.

Chapter 12 Practice questions

1   Which of the following can be used in a vaccine?

    1   An inactivated virus

    2   Antibodies specific to a bacterial pathogen

    3   A protein fragment from a virus

    A   1, 2, and 3 are correct

    B   Only 1 and 3 are correct

    C   Only 2 and 3 are correct

    D   Only 1 is correct   *(1 mark)*

2   Which of the following statements best describes how herd immunity is usually achieved?

    A   A large proportion of a population survives a disease epidemic.

    B   A large proportion of a population is vaccinated.

    C   A large proportion of a population is given antibodies specific to a pathogen.

    D   A large proportion of a population receives antibodies specific to a pathogen from their mothers during gestation.   *(1 mark)*

3   Which of the following statements describes the typical mechanism of bacterial pathogenicity?

    A   Integration of genetic material into host cell chromosomes

    B   Enzyme secretion

    C   Production of toxins

    D   Activation of host cell genes   *(1 mark)*

4   A nation has a population of 3.4×10^7 people.

    In one year, this nation had 2.0×10^5 cases of tuberculosis (TB), of which 40 000 were new cases.

    a   What was the incidence rate (per 100 000) of TB in this nation?   *(1 mark)*

    b   The same nation had 6.8×10^3 deaths from TB over a one-year period.

       What was the mortality rate (per 100 000) from TB in this nation?   *(1 mark)*

5   Suggest why attempts to develop a vaccine against HIV have been unsuccessful.   *(2 marks)*

6   **a**   Outline methods farmers can use to reduce the spread of disease between crop plants.   *(3 marks)*

    b   Outline methods that hospital staff can use to reduce the spread of disease within hospitals.   *(2 marks)*

7   Complete the following passage about plant responses to pathogens. Choose the most appropriate word to place in each gap.   *(5 marks)*

    Pathogenic molecules can bind to .... in the cell membranes of plant cells. This triggers a cascade of reactions that results in genes in the .... being switched on. A polysaccharide called .... is produced, which is deposited in cell walls and .... sieve plates. These physical barriers are further strengthened by ....

13.1 Coordination

Specification reference: 5.1.1 (a) and (b)

Synoptic link

The remainder of this chapter is concerned with the nervous system. You will learn about two other examples of communication systems in Chapter 14, Hormonal communication, and Chapter 16, Plant responses.

Revision tip: A wonderful reception

The ideas of binding specificity and complementary shapes are relevant to discussions of both enzymes and cell receptors. However, when discussing signalling molecules and cell receptors you should use the term 'binding site'. Reserve the term 'active site' for enzymes.

Multicellular organisms have evolved several specialised systems for responding to internal and external stimuli. These different systems need to be coordinated, which requires cell-to-cell communication (**cell signalling**).

Communication systems

Communication systems involve cells releasing chemicals that stimulate responses in other cells. This is known as cell signalling. **Homeostasis**, which you will read about in Chapter 15, relies on cell signalling between different organs in order to produce coordinated responses.

Cell signalling requires a cell to release a signalling molecule, which binds to **receptors** on a **target cell**.

Examples of communication systems

Communication system	Topic references	How do cells signal to each other?
Nervous system	13.2–13.9	Neurotransmitters (signalling molecules) diffuse from a presynaptic neurone and bind to receptors on postsynaptic neurones.
Animal endocrine system	14.1	Hormones (signalling molecules) are released from glands, travel through blood, and bind to receptors on target cells.
Plant hormonal communication	Chapter 16	Hormones (signalling molecules) diffuse through plant tissue to target cells.

Go further: Types of cell signalling

Types of cell signalling can be classified based on the destination of the signalling molecule.

Autocrine signalling: the cell that produces the signalling molecule is also the target cell.

Juxtacrine signalling: the target cells are adjacent to the cells producing the signalling molecule.

Paracrine signalling: target cells are nearby the cell that produces the signalling molecule.

Endocrine signalling: target cells are usually not near the producing cells.

1 Suggest, with a reason, the form of cell signalling used by **a** neurones **b** pancreatic β cells.

Summary questions

1 Complete the passage below by placing the most appropriate words or phrases in the gaps.
In the nervous system, ... act as signalling molecules for neurone-to-neurone communication. The .... neurone is the target cell. In the .... system, hormones act as signalling molecules. *(3 marks)*

2 Describe the role of cell signalling in the nervous system. *(2 marks)*

3 Explain why homeostasis relies on cell signalling. *(2 marks)*

13.2 Neurones

Specification reference: 5.1.3 (b)

The nervous system comprises specialised cells called **neurones**. The electrical impulses transmitted by neurones enable communication between cells in different parts of an organism. Several types of neurone exist.

Neurone structure

A nervous response usually follows this pathway:

Sensory receptor → sensory neurone → relay neurone → motor neurone → effector

▼ **Table 1** *Features of the three types of neurone*

Type of neurone	Function	Key structural features
Sensory	Transmits impulses from receptors to relay neurones in the central nervous system	One **dendron** (carrying the impulse from the receptor to the cell body). One **axon** (carrying the impulse from the cell body to a relay neurone).
Relay	Transmits impulses between neurones	Many clusters of **dendrites**, each leading to a dendron. Each dendron passes to the **central cell body**. A short axon carries impulses from the cell body to many synaptic endings.
Motor	Transmits impulses from a relay neurone to an effector (i.e. muscle or gland)	Dendrites leading to the cell body. One **long axon** (carrying impulses from cell body to effector).

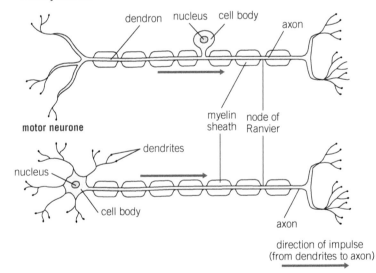

sensory neurone

dendron   nucleus   cell body   axon

myelin sheath   node of Ranvier

motor neurone

dendrites

nucleus

cell body

axon

direction of impulse
(from dendrites to axon)

▲ **Figure 1** *Sensory and motor neurones*

Myelinated vs non-myelinated

The axons of motor neurones, many sensory neurones, and some relay neurones are covered in **myelin sheath**, which is a type of fat produced by **Schwann cells**. Myelin speeds up nervous impulses by enabling **saltatory conduction** (see Topic 13.4, Nervous transmission).

Revision tip: Schwann song?
You may be asked to identify myelin sheath and Schwann cells from photographs. Myelin sheath can be seen as a thick band surrounding the axon of a neurone. The Schwann cell will appear as an irregular shape and will be located outside the band of myelin sheath.

Synoptic link
You will learn about action potentials (Topic 13.4, Nervous transmission) and synapses (Topic 13.5, Synapses) later in this chapter.

Revision tip: What a nerve!
Neurones are individual cells. **Nerves** consist of many neurones bundled together.

Revision tip: Cell bodies
The cell body contains a neurone's nucleus and rough ER. This is where new proteins are produced within a neurone.

Revision tip: Is that relay the correct name?
Relay neurones are also known as **interneurones** and **intermediate neurones**.

Summary questions

1   State two structural differences between a sensory neurone and a motor neurone.   (*2 marks*)

2   Many relay neurones are not surrounded by myelin sheath. Suggest why.   (*2 marks*)

3   Motor neurones can be over one metre long in humans. The cell body of a motor neurone is located at one end of the cell.
    a   Suggest one problem that this structural arrangement presents for a motor neurone.   (*2 marks*)
    b   Suggest which features may be present in the neurone to overcome this problem.   (*1 mark*)

13.3 Sensory receptors

Specification reference: 5.1.3 (a)

Receptors detect changes in the environment and convert the stimulus into an electrical impulse. Stimuli can take many forms, and they can be either internal or external, which has resulted in the evolution of a wide range of receptor structures.

Types of receptor

Receptors are located in sense organs (e.g. the eyes, ears, nose, tongue, and skin). They convert a stimulus to a nervous impulse, which is transmitted along a sensory neurone.

▼ **Table 1** *Types of sensory receptor*

Stimulus	Receptor	Mechanistic detail
Pressure	Pacinian corpuscle	Mechanical pressure applied to the skin opens stretch-mediated sodium ion channels, triggering an action potential (see Topic 13.4, Nervous transmission) in a sensory neurone.
Light	Rods cells (in the retina of the eye)	Light causes a chemical reaction to occur in the rod cells (i.e. the breakdown of rhodopsin), which alters the permeability of the cell membrane to sodium ions.
Chemicals	Taste receptor / Olfactory cells	Molecules or ions (e.g. sugars, salt, odour molecules) bind to receptors on the receptor cell membranes. This causes a second messenger response, similar to the response produced by hormones (see Topic 14.1, Hormonal communication). cAMP levels rise and alter the permeability of the cell membranes to Na^+ ions. Depolarisation occurs and triggers an action potential.
Temperature	Thermoreceptors	Specialised sensory neurones. The permeability of their membranes to Na^+ ions changes with temperature.
Sound	Hair cells (in the ear)	Sound waves move cilia on hair cells, which triggers changes in membrane permeability to K^+ ions.

Summary questions

1 Explain why sensory receptors are considered to be transducers.
(1 mark)

2 State the form of energy converted to electrical impulses by
    a rod cells **b** thermoreceptors **c** Pacinian corpuscles
    d olfactory cells.
(4 marks)

3 cAMP is generated by the binding of hormones to target cells and, in sensory receptors, in response to certain stimuli. Using the information in Table 1, describe the difference in the action of cAMP in these two types of cell.
(3 marks)

Impulses are transmitted through a neurone by temporary changes in the voltage across the neurone's cell membrane. This change in voltage is called an **action potential**. When no impulses are being transmitted, the voltage across a neurone's membrane is known as the **resting potential**.

Resting potential

The cell membrane of a neurone is **polarised** when not firing impulses (i.e. the extracellular fluid outside the neurone is positively charged, and the cytoplasm in the neurone is negatively charged). The potential difference across the membrane (approximately **–70 mV**) is known as the resting potential. The distribution of two ions (sodium (Na^+) and potassium (K^+)) determines the resting potential.

▼ **Table 1** *How is the resting potential established?*

Membrane protein	How does it produce the resting potential?
Sodium–potassium pump	Three Na^+ ions are pumped out of the neurone (by active transport) for every two K^+ ions pumped in. This sets up an imbalance of positive charge (i.e. outside the neurone is more positive than inside).
K^+ channels	Whereas Na^+ channels are closed at the resting potential, some K^+ channels remain open. This enables some K^+ ions to diffuse out of neurone, down a concentration gradient. Even more positive charge therefore builds up outside the neurone.

Revision tip: In the balance...

The concentration gradient of sodium ions across an axon membrane is steeper than the potassium ion concentration gradient.

OUTSIDE

Na⁺   K⁺   Axon membrane

Na⁺   K⁺   INSIDE

This is why overall the inside of an axon at rest has fewer positive ions than the outside.

Negative ions are also involved in establishing the resting potential, but you are not required to learn about these ions.

Common misconception: Potassium channels

Two forms of potassium ion (K^+) channel proteins are present in neuronal cell membranes.

Some K^+ channels remain open constantly and are not voltage-gated. These channels help establish the resting potential.

Some K^+ channels are voltage-gated. These open during an action potential (stage ④ in Figure 1) in response to an influx of positive charge.

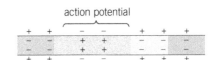

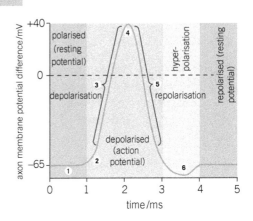

▲ **Figure 1** *Changes in potential difference during an action potential*

Action potential

The detection of a stimulus by a receptor (see Topic 13.3, Sensory receptors) initiates a change in voltage across a neurone's membrane. The potential difference switches to approximately +40 mV (i.e. the inside becomes more positive than the outside of the neurone). The membrane is **depolarised**. This is known as an action potential.

What happens during an action potential?

1 The **resting potential** (see stage ① in Figure 1).
2 **Na^+ channels open**. Na^+ ions diffuse into the neurone down an **electrochemical gradient**.

Key terms

Resting potential: The potential difference of a neurone membrane at rest (approximately −65 mV).

Action potential: The change in potential difference across a neurone membrane following a stimulus (approximately +40 mV).

Synoptic link

The propagation of a nerve impulse is an example of positive feedback, which you will learn about in Topic 15.1, The principles of homeostasis.

Key terms

Depolarisation: The change in potential difference across a neurone membrane from negative to positive (e.g. during an action potential).

Repolarisation: The change in potential difference across a neurone membrane from positive to negative (e.g. the restoration of the resting potential).

Hyperpolarisation: The overshoot of a neurone membrane's potential difference following an action potential; it becomes more negative than the resting potential.

3 The initial influx of Na⁺ ions causes more voltage-gated Na⁺ channels to open (**depolarisation**).

4 Na⁺ ions continue to diffuse into the neurone through voltage-gated Na⁺ channels until the potential difference reaches +40 mV. The voltage-gated **Na⁺ channels then close** and voltage-gated **K⁺ channels open**.

5 K⁺ ions diffuse out of the neurone, reducing the positive charge inside the neurone and **repolarising** the membrane.

6 Voltage-gated K⁺ channels close. The membrane becomes **hyperpolarised** (i.e. due to the K⁺ ions leaving, the inside of the neurone becomes more negative than it is in the resting state). This is known as a **refractory period** – no more action potentials can occur until the resting potential is restored. Sodium–potassium pumps return the neurone to its resting potential.

Propagation

An action potential is propagated (i.e. spread) along the length of a neurone; this wave of depolarisation is called a **nerve impulse**. The propagation of an action potential involves the following events:

1 Sodium ions enter a neurone and depolarise it.

2 The sodium ions diffuse further along the neurone.

3 The increased positive charge caused by the diffusion of sodium ions opens more (neighbouring) voltage-gated sodium ion channels.

4 The action potential passes along the neurone.

Speed of impulses

Three factors increase the speed of nerve impulses:

- A greater **axon diameter** (which reduces resistance to ion flow).
- A higher **temperature**.
- The presence of **myelin**.

Neurones that are covered in myelin can transmit impulses at a faster rate. Impulses travel faster due to **saltatory conduction** – i.e. the axon membrane can depolarise only at gaps in the myelin (known as **nodes of Ranvier**). The action potential effectively jumps between nodes, which is more efficient than the entire membrane being depolarised.

Summary questions

1 Complete the following paragraph by adding the most appropriate words or phrases to the gaps. *(4 marks)*
A neurone's resting potential is set up by sodium–potassium pumps, which transfer .... Na⁺ ions to the outside of the neurone for every .... K⁺ ions transferred inside. During an action potential, the neurone first becomes .... when Na⁺ ions diffuse into the axon. The subsequent outward movement of K⁺ ions results in ....

2 Describe how the transmission of nerve impulses differs between myelinated and unmyelinated neurones. *(3 marks)*

3 Explain why a neurone will not produce action potentials that vary in magnitude. *(3 marks)*

13.5 Synapses

Specification reference: 5.1.3 (d)

Once a nervous impulse reaches the end of a neurone, the signal is transmitted to another neurone across a synapse. This is an example of cell signalling.

Synapse structure and transmission

Transmission across a synapse involves the following steps:

1 An action potential passes to the end of the **presynaptic neurone** (presynaptic knob).

2 Voltage-gated **Ca^{2+} ion channels open.**

3 A Ca^{2+} ion influx causes **vesicles** containing neurotransmitters to fuse with the neurone's cell membrane.

4 **Neurotransmitters** are released into the **synaptic cleft** by **exocytosis**.

5 The neurotransmitters **diffuse** across the synaptic cleft and bind to **receptors** on the **postsynaptic neurone's** cell membrane.

6 **Sodium ion channels** are triggered to open, causing depolarisation and an action potential in the postsynaptic neurone (if sufficient neurotransmitters bind and the threshold value is surpassed).

Synapse roles

Synapses play the following roles in the nervous system:

- Ensure impulses travel in only **one direction** (because neurotransmitters are released only from presynaptic neurones, and receptors are found only on postsynaptic neurones).

- One presynaptic neurone can signal to many postsynaptic neurones. This enables signals to be passed to different effectors.

- Enable one sensory neurone to receive signals from several receptors. This provides information about the extent of the stimulus. **Spatial summation** occurs – i.e. sometimes an action potential in a postsynaptic neurone will occur only if several presynaptic neurones release neurotransmitters.

- Enable **temporal summation** – i.e. sometimes a postsynaptic action potential occurs only when several impulses have travelled down a presynaptic neurone. Each impulse releases more neurotransmitters until the threshold depolarisation is surpassed in the postsynaptic neurone.

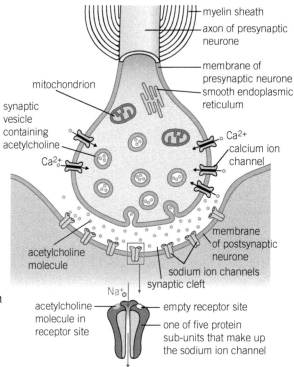

▲ **Figure 1** *Key features of a synapse (represented here by a cholinergic synapse)*

> **Revision tip: Excitement or inhibition?**
> Several neurotransmitters exist. Some (such as acetylcholine) are **excitatory** – i.e. produce action potentials in postsynaptic neurones. Others are **inhibitory** – i.e. they hyperpolarise postsynaptic membranes, reducing the possibility of an action potential.

> **Revision tip: Cleft and right?**
> A **synaptic cleft** is specifically the gap between a presynaptic and postsynaptic neurone. The term 'synapse' refers to the combination of the presynaptic membrane, the cleft, and the postsynaptic membrane. You should write that 'neurotransmitters diffuse across the synaptic cleft' (rather than the synapse).

> **Revision tip: Recycling is important**
> In most cases, neurotransmitters are reabsorbed back into the presynaptic neurone once they have performed their cell signalling function. **Acetylcholine**, for example, is broken down by an enzyme called **acetylcholinesterase** (found on the postsynaptic membrane). The products (choline and ethanoic acid) are reabsorbed across the presynaptic membrane.

Summary questions

1 Describe how acetylcholine is released from a presynaptic neurone.
(3 marks)

2 Use the concept of temporal summation to explain why a weak stimulus may be filtered out and not produce a response from the nervous system.
(3 marks)

3 Suggest why it is important for neurotransmitters such as acetylcholine to be recycled.
(2 marks)

13.6 Organisation of the nervous system
13.7 Structure and function of the brain

Specification reference: 5.1.5 (h) and 5.1.5 (g)

The mammalian nervous system is organised into neurones with a coordinating role (the central nervous system) and neurones carrying signals to and from the rest of the body (the peripheral nervous system). Neurones within these two regions can be further categorised based on their roles.

Nervous system organisation

The divisions of the nervous system are illustrated in Figure 1.

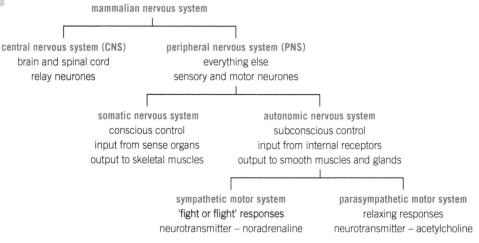

▲ **Figure 1** *Organisation of the mammalian nervous system*

Structure of the brain

The billions of neurones in the brain are organised into regions with different functions.

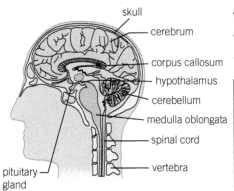

▲ **Figure 2** *The gross structure of the brain*

▼ **Table 1** *The functions of some regions of the brain*

Region of the brain	Function
Cerebrum	Coordinates **voluntary** responses
Cerebellum	Controls **balance** and posture
Medulla oblongata	**Autonomic** functions (e.g. heart rate and breathing rate)
Hypothalamus	**Autonomic** functions (e.g. thermoregulation)
Pituitary gland	Releases **hormones** that control other glands in the body

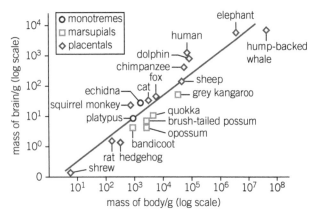

▲ **Figure 3**

Summary questions

1   Which specific branch of the peripheral nervous system would produce the following responses: **a** a person gripping a cup with their hand **b** an increase in sweat production **c** constriction of pupils.   *(3 marks)*

2   Explain why a person struggling to move normally may have suffered damage to either their cerebrum or cerebellum.   *(2 marks)*

3   What do the data in Figure 3 suggest about the evolution of the human brain?   *(2 marks)*

13.8 Reflexes

Specification reference: 5.1.5 (i)

A reflex is an involuntary response to a stimulus. They are faster than responses that require conscious thought; such rapid responses can protect the body from danger.

The reflex arc

Rather than using the cerebrum for complex processing, a reflex relies on a simple pathway:

Sensory receptor → sensory neurone → relay neurone → motor neurone → effector

The relay neurones in most reflexes are found in the spinal cord, but some are located in the lower part of the brain.

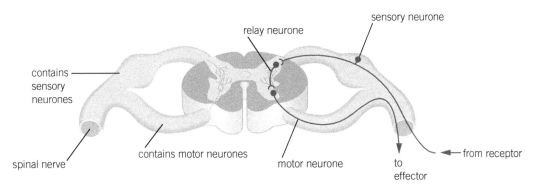

▲ **Figure 1** *A reflex arc*

Examples of reflexes

You need to have an understanding of two reflexes in particular: the knee-jerk reflex and the blinking reflex.

▼ **Table 1** *Two examples of reflexes*

	Knee-jerk reflex	Blinking reflex
Stimulus	Firm tap below the kneecap	Touch on the cornea
Receptor	Stretch receptor in muscle	Touch receptors in the cornea
Location of relay neurone	Spinal cord	Lower brain stem
Effector	Muscles in the upper leg	Muscles in the eyelids
Importance	Maintaining balance	Preventing damage to the eyes

Summary questions

1   State the location of the relay neurones in
    a the knee-jerk reflex b the blinking reflex.   (*2 marks*)

2   Describe and explain the general characteristics of reflex responses
    that aid the survival of organisms.   (*4 marks*)

3   Explain how the typical neural pathway of a reflex maximises its
    effectiveness.   (*3 marks*)

Revision tip: Survival instincts

Reflexes help organisms to survive because they are:

- very **fast** (with only two synapses)

- **innate** (not requiring learning)

- **involuntary** (which frees the brain to process more complex decision-making, if required).

13.9 Voluntary and involuntary muscles
13.10 Sliding filament model
Specification reference: 5.1.5 (I)

A muscle is an example of an **effector**. Impulses transmitted by motor neurones stimulate muscle cells to contract and produce a response.

Types of muscle

Muscles can be either under conscious control (i.e. voluntary, **skeletal** muscle) or under the control of the autonomic nervous system (i.e. involuntary muscle, which can be either **smooth** or **cardiac**).

▼ **Table 1** Types of muscle

Type of muscle:	Skeletal	Cardiac	Smooth (involuntary)
Appearance	Striated (i.e. striped)	Striated (but with fainter striations than skeletal muscle)	Non-striated
Location	Attached to bones via tendons	Heart	Walls of blood vessels, digestive system, and excretory system
Type of contraction	Voluntary (conscious) Fast Short in duration	Involuntary Intermediate speed Intermediate duration	Involuntary Slow Can be long-lasting

Skeletal muscle structure

Muscle cells fuse to form fibres. Each **muscle fibre** contains many **myofibrils**, which are organelles made principally of two proteins: **actin** and **myosin**. Myofibrils are composed of many repeating units called **sarcomeres**.

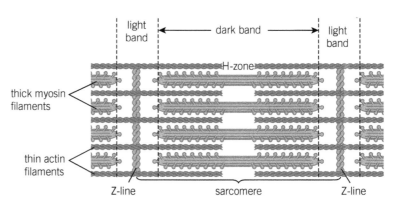

▲ **Figure 1** A sarcomere

Muscle contraction

A muscle contracts when a signal is received from a motor neurone, causing sarcomeres to contract in unison.

Neuromuscular junctions

A **neuromuscular junction** is the synapse between a motor neurone and a muscle fibre. It works using the same principles as a synapse between two neurones: a neurotransmitter (acetylcholine) diffuses across the synaptic cleft and binds to receptors on the sarcolemma, resulting in depolarisation.

A **motor unit** comprises all the muscle fibres supplied by one motor neurone.

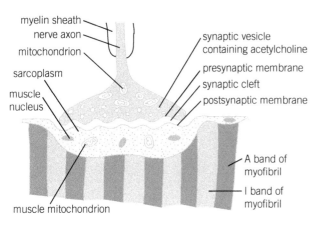

▲ **Figure 2** *A neuromuscular junction*

Sliding filament model

Muscle contraction involves the following steps, which represent the sliding filament model:

1 The sarcolemma is depolarised.

2 The depolarisation spreads through **T-tubules** to **sarcoplasmic reticulum** (specialised smooth ER).

3 **Ca²⁺ ions** are released from the sarcoplasmic reticulum.

4 Ca²⁺ ions bind to **troponin** (a protein that is attached to **actin**).

5 The **troponin changes shape**, which causes **tropomyosin** (another protein) to be moved away from the myosin binding site it had been covering.

6 **Myosin heads** bind to the binding sites on actin. This forms **cross bridges**.

7 Myosin **heads tilt**, thereby moving the actin. This is called the **power stroke**. **ADP is released** from myosin at this stage.

8 **ATP binds** to myosin, causing it to detach from the actin.

9 **ATP is hydrolysed** to ADP, causing the myosin head to resume its original position. The head is free to attach further down the actin. More than 100 power strokes can be performed by each myosin every second.

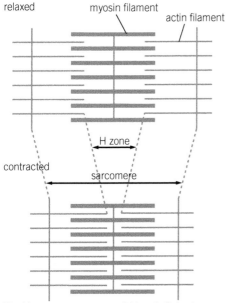

The H zone, sarcomere, and I band all shorten.
The A band is unaltered

▲ **Figure 3** *Changes in a sarcomere during muscle contraction*

Summary questions

1 How will the length of the following sarcomere features change during muscle contraction:
   a A band
   b I band
   c myosin filaments
   d actin filaments
   e H zone.   (*5 marks*)

2 Outline the role of calcium ions in muscle contraction.   (*3 marks*)

3 Suggest how the ultrastructure of a muscle fibre differs from that of an epithelial cell. Explain the reasons for these differences.   (*6 marks*)

1 Which of the following statements is/are true of the functions of the sympathetic nervous system? *(1 mark)*

   1 Increases heart rate.

   2 Increases the speed at which food moves through the gut.

   3 Decreases sweat production.

   A 1, 2, and 3 are correct

   B Only 1 and 2 are correct

   C Only 2 and 3 are correct

   D Only 1 is correct

2 What is the function of the cerebellum? *(1 mark)*

   A Coordination of balance and muscular movement.

   B Controls conscious thought processes.

   C Controls body temperature.

   D Regulates heart rate.

3 Which of the following is/are present in the membrane of an acetylcholinergic presynaptic neurone? *(1 mark)*

   1 Calcium ion channel

   2 Acetylcholine receptor

   3 Acetylcholinesterase

   A 1, 2, and 3 are correct

   B Only 1 and 2 are correct

   C Only 2 and 3 are correct

   D Only 1 is correct

4 Which of the following statements is/are true of the role of calcium ions in skeletal muscle contraction? *(1 mark)*

   1 Calcium ions are released into the sarcoplasm.

   2 Calcium ions bind to tropomyosin.

   3 Calcium ions inactivate an enzyme called myosin kinase.

   A 1, 2, and 3 are correct

   B Only 1 and 2 are correct

   C Only 2 and 3 are correct

   D Only 1 is correct

5 Which of the following statements is/are true of the changes in a sarcomere during muscle contraction? *(1 mark)*

   1 Myosin filaments shorten.

   2 The H zone shortens.

   3 I band shortens.

   A 1, 2, and 3 are correct

   B Only 1 and 2 are correct

   C Only 2 and 3 are correct

   D Only 1 is correct

6 Outline the differences between the sympathetic nervous system and the parasympathetic nervous system. *(6 marks)*

14.1 Hormonal communication

Specification reference: 5.1.4 (a) and (b)

You learned about one biological communication system, the nervous system, in the previous chapter. Here you will learn about the endocrine system, which uses chemical messengers called hormones to enable communication between cells.

The endocrine system

Endocrine **glands** secrete chemicals called **hormones** into the circulatory system. Hormones are transported in the **blood** to **target cells**.

Types of hormones

Hormones produce a response in target cells. However, the method by which they produce a response depends on the type of hormone.

Type of hormone	Chemistry	How does it affect target cells?	Examples
Steroid	Lipid-soluble	Diffuses through the cell surface membrane Binds to a receptor (in cytoplasm or nucleus) Promotes or inhibits transcription (of a particular gene)	Testosterone Oestrogen Glucocorticoids Mineralocorticoids
Non-steroid	Hydrophilic (e.g. polypeptides or glycoproteins)	Binds to a receptor on the cell surface membrane Activates a second messenger (e.g. cyclic AMP) A cascade of intracellular reactions activates a transcription factor (for a particular gene)	Glucagon Insulin Adrenaline

The adrenal glands

The adrenal glands produce both steroid hormones (from the cortex region) and non-steroid hormones (from the medulla).

Region of adrenal glands	Hormone	Function
Cortex	Glucocorticoids	Regulates carbohydrate metabolism
	Mineralocorticoids	Regulates salt and water concentrations
	Androgens (e.g. testosterone)	Various functions (e.g. regulation of muscle mass, development of sexual characteristics)
Medulla	Adrenaline	Increases heart rate and blood glucose concentration (by stimulating glycogenolysis)
	Noradrenaline	Works in concert with adrenaline by increasing heart rate, widening airways, and increasing blood pressure

Key terms

Endocrine gland: An organ that secretes hormones.

Hormone: A chemical messenger that is transported in the blood.

Revision tip: First things first

Hormones can be referred to as **first messengers**. Hormones stimulate the production of **second messengers** (e.g. cyclic AMP) within target cells.

Summary questions

1 Define **a** a target cell **b** a second messenger. *(3 marks)*

2 Outline the similarities and differences between the modes of action of steroid and non-steroid hormones. *(4 marks)*

3 Suggest how the adrenal glands respond to physiological changes in an athlete running a marathon. *(5 marks)*

14.2 The pancreas

Specification reference: 5.1.4 (c)

The pancreas is an organ that has both **endocrine** functions (i.e. hormone production) and **exocrine** functions (i.e. enzyme production).

Pancreatic functions

Role of the pancreas	What is produced?	Where is it produced?
Exocrine gland	Amylase	These digestive enzymes are secreted from **exocrine tissue** (which constitutes most of the pancreas) and are released into the **pancreatic duct**, which leads to the duodenum (the upper small intestine).
	Proteases	
	Lipases	
Endocrine gland	Insulin	β (beta) cells (in the islets of Langerhans).
	Glucagon	α (alpha) cells (in the islets of Langerhans).

Pancreatic structure

The pancreas contains two types of cell (the **islets of Langerhans** and exocrine cell clusters called **acini**), which can be distinguished through staining.

Structure	How can you identify them?
Islets of Langerhans (endocrine)	A large cluster of cells that is usually stained **blue/lilac** in photographs
Acini (exocrine)	Small clusters of cells stained **dark pink/purple**

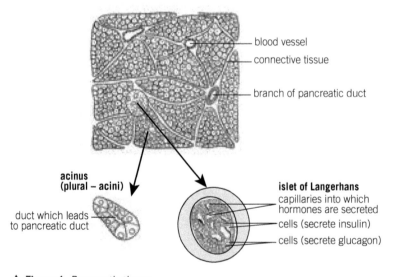

blood vessel

connective tissue

branch of pancreatic duct

acinus (plural – acini)

duct which leads to pancreatic duct

islet of Langerhans
capillaries into which hormones are secreted
cells (secrete insulin)
cells (secrete glucagon)

▲ **Figure 1** *Pancreatic tissue*

> ### Revision tip: Exocrine or endocrine?
> **Endocrine** glands secrete **hormones** directly into the **blood. Exocrine** glands secrete substances (often **enzymes**) into **ducts**.

> ### Synoptic link
> You may be required to examine and draw pancreatic tissue under a microscope. Remind yourself of the necessary differential staining and microscopy skills by reading Topic 2.1, Microscopy.

> ### Revision tip: Ducts and vessels
> You may be asked to identify ducts and blood vessels within pancreatic tissue. These are not always easy to differentiate. As a rule of thumb, branches of the pancreatic duct will tend to have thick walls. Blood cells may sometimes be observable within blood vessels.

Summary questions

1 Describe how cells in the islets of Langerhans can be distinguished from pancreatic acini cells.   (*2 marks*)

2 Explain the difference between exocrine and endocrine glands, using the pancreas as an example.   (*4 marks*)

3 Suggest how transmembrane transport of ions and molecules across the cell membranes of pancreatic α and β cells will differ.   (*5 marks*)

14.3 Regulation of blood glucose concentration

Specification reference: 5.1.4 (d)

The regulation of blood glucose is coordinated by endocrine cells in the pancreas and is an example of homeostasis.

How does the pancreas regulate blood glucose concentration?

	When blood glucose needs to be increased	When blood glucose needs to be decreased
What is secreted?	Glucagon (from α cells)	Insulin (from β cells)
Where does it have an effect?	Binds to receptors on liver cells	Binds to receptors on liver and muscle cells
What effects are produced?	Less glucose taken up by cells More fatty acids are used in respiration Glycogen converted to glucose (**glycogenolysis**) Amino acids and fats converted to glucose (**gluconeogenesis**)	More glucose absorbed by cells Glucose is converted to fats More glucose used in respiration Glucose converted to glycogen (**glycogenesis**) Inhibits glucagon release

Insulin secretion

Insulin is secreted from β cells using the following mechanism:

1 Blood glucose concentration is high. **2 Glucose enters** a β cell through a **transporter** protein in the cell surface membrane. **3** Glucose is metabolised to produce **ATP. 4** ATP binds to and closes **K⁺ ion channels** in the cell surface membrane. **5** K⁺ ions build up in the cell and increase the positive charge in the cell (**depolarisation**). **6** Voltage-gated **Ca²⁺ ion channels** open. **7** Ca²⁺ ions diffuse into the cell and trigger secretory **vesicles** to release **insulin** from the cell via **exocytosis**.

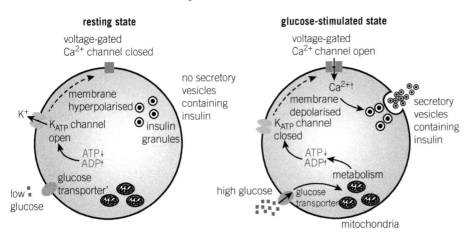

▲ **Figure 1** *The mechanism of insulin release in beta cells*

Revision tip: (Negative) feedback is welcome

Insulin and glucagon have antagonistic effects. The lowering of blood glucose by insulin and the raising of blood glucose by glucagon are examples of **negative feedback**.

Synoptic link

You learned about the structure of glucose in Topic 3.3, Carbohydrates. In Topic 15.1, The principles of homeostasis, we will discuss the concept of negative feedback.

Revision tip: G is for... ?

Many of the terms in this topic begin with 'g'. Understanding the following is important:

'lysis' = splitting
'genesis' = creation/formation
'neo' = new

Therefore:

glycogenesis = the formation of glycogen
glycogenolysis = the splitting of glycogen
gluconeogenesis = the formation of new glucose

Summary questions

1 Describe the role of ATP in the release of insulin from beta cells in the pancreas.
    (2 marks)

2 The set point for blood glucose concentration in humans is approximately 90 mg 100 cm⁻³. How would this value be expressed in g cm⁻³? *(1 mark)*

3 Suggest why liver cells have specialised receptors for glucagon. *(3 marks)*

14.4 Diabetes and its control

Specification reference: 5.1.4 (e) and (f)

Diabetes mellitus is a disease in which the homeostatic control of blood glucose concentration is lost. Untreated, the level of glucose in the blood of a diabetic patient is liable to be too high.

Types of diabetes

Characteristic	Type 1 (insulin dependent)	Type 2 (insulin independent)
Insulin production	Little or none	Often reduced
Cause	Usually an autoimmune response (β cell destruction, which stops insulin production)	Effector cells lose responsiveness to insulin
Genes vs environment?	Genetic	Genetic and environmental (e.g. diet and activity levels) factors influence the risk of developing the disease
Age at onset	Childhood (juvenile-onset)	Adulthood (late-onset)
Speed of onset	Quick	Slow
Usual treatments	Insulin injections	Dietary control of carbohydrate intake

 Go further: Maturity-onset diabetes of the young

Forms of diabetes mellitus exist other than type 1 and 2, although they are rare. Maturity-onset diabetes of the young (MODY) is a hereditary condition that constitutes 2% of diabetes mellitus cases. It shows autosomal dominant inheritance (you can remind yourself of inheritance patterns by reading Topic 20.2, Monogenic inheritance).

MODY tends to be diagnosed in people below 30 years of age. It is usually characterised by mild hyperglycaemia (high blood glucose concentration). MODY can be controlled in the early stages by planning meals and may not require insulin injections. Patients are not insulin resistant; the problem lies with glucose metabolism or insulin secretion. Several different forms of MODY have been identified, each caused by a genetic variant at a single gene locus.

1 Suggest why MODY is sometimes referred to as 'monogenic diabetes'.

2 What is the range of probability that two parents with MODY will have a child with the condition?

3 MODY patients with heterozygous genotypes can often manage their condition through careful dietary planning. Suggest why MODY patients with homozygous genotypes are likely to require insulin injections.

Synoptic link

In Topic 21.4, Genetic engineering, you will learn more about the process of genetically modifying bacteria to produce insulin and other proteins. You learned about stem cells in Topic 6.5, Stem cells.

Question and model answer: Insulin sources

Q. Describe how the sources of insulin for type 1 diabetes treatment are changing, and explain the benefits of these changes.

A. For many years, insulin was obtained from the pancreas of other animals (e.g. pigs).

The production of human insulin was made possible by using genetically modified (GM) bacteria. This method has some advantages:

- It's cheaper than extracting non-human insulin from other animals.

- The rate of production is higher.

> When answering these type of questions, ensure you link each description to at least one explanation.

More recently, some patients have been injected with β cells. Stem cell therapy is being researched as an alternative treatment. This has advantages over GM insulin production and β cell transplant:

- Patients would no longer require insulin injections.

- There would be a low risk of rejection.

Summary questions

1 Why is type 1 diabetes sometimes called insulin-dependent and type 2 called insulin-independent? *(2 marks)*

2 Study Figure 1. Describe and explain the differences in the responses of diabetics and non-diabetics following the consumption of glucose. *(3 marks)*

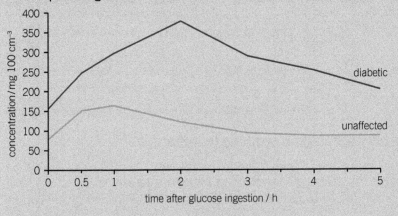

▲ **Figure 1**

3 Suggest why the prevalence of type 2 diabetes is likely to have a greater impact on future populations than the prevalence of type 1 diabetes. *(5 marks)*

14.5 Coordinated responses

Specification reference: 5.1.5 (j)

The nervous and endocrine (hormonal) systems often work together in response to stimuli. The **fight or flight** response in mammals is an example of such a coordinated response.

Endocrine and nervous system responses

▼ **Table 1** *A comparison of the endocrine and nervous systems*

	Endocrine (hormonal) system	Nervous system
What is transmitted?	Chemicals (hormones) in the blood	Electrical impulses along neurones
How quick is the communication?	Relatively slow	Rapid
For how long do the effects last?	Long-lasting and can be permanent	Short-lived and temporary

Fight or flight response

Mammals react to danger through the fight or flight response. This response is coordinated by the **sympathetic** branch of the **autonomic nervous system** (and is controlled by the **hypothalamus**). However, hormonal communication also plays a role; the sympathetic nervous system stimulates the release of many hormones, including **adrenaline** (from the adrenal medulla) and **cortisol** (a stress hormone produced in the adrenal cortex).

▼ **Table 2** *Fight or flight responses*

Physical response		How does this benefit the mammal?
Increase in…	Heart rate	Oxygen and glucose are circulated faster
	Blood glucose concentration	Respiration rate in cells is raised
	Pupil dilation	Improves vision
Decrease in…	Blood flow to skin surface	More blood can be diverted to skeletal muscle, the brain, and the heart
	Digestion rate	
	Concentration on small tasks	The brain focuses on the immediate threat

How does adrenaline produce its effects?

Adrenaline is a non-steroid hydrophilic hormone. You read about adrenaline and the action of non-steroid hormones in Topic 14.1, Hormonal communication. Adrenaline produces effects in the target cell using the following mechanism:

- It binds to a cell surface **receptor**.
- It activates **adenylyl cyclase** (which is an enzyme).
- Adenylyl cyclase converts ATP to cyclic AMP (**cAMP**) (a second messenger).
- cAMP activates protein kinases, which activate other enzymes (this is known as a **cascade effect**).

Synoptic link

Chapter 13 focused on the nervous system, and you learned about the structure and function of the adrenal glands in Topic 14.1, Hormonal communication.

Summary questions

1 Describe the role of adenylyl cyclase in a liver cell. *(2 marks)*

2 Explain why the fight or flight response is considered a coordinated response (i.e. controlled by both the nervous system and the endocrine system). *(2 marks)*

3 Suggest why a mammal responding to danger may experience a loss of hearing. *(2 marks)*

14.6 Controlling heart rate

Specification reference: 5.1.5 (k)

Heart rate is controlled by the autonomic nervous system and, in times of stress, hormones.

How is heart rate controlled?

Changes in heart rate are controlled by the cardiovascular centre in a region of the brain called the **medulla oblongata**.

▼ **Table 1** How changes to heart rate are initiated by the medulla oblongata

	To increase heart rate	To decrease heart rate
Which branch of the autonomic nervous system is used?	Sympathetic	Parasympathetic
Which nerve transmits the impulse from the medulla oblongata?	Accelerator nerve	Vagus nerve
What effect is produced in the heart?	Increases the rate at which the SAN generates impulses	Decreases the rate at which the SAN generates impulses

Role of receptors

Two different types of receptor pass information to the medulla oblongata: chemoreceptors and pressure receptors (baroreceptors).

▼ **Table 2** The roles of receptors in the regulation of heart rate

	Chemoreceptors	Baroreceptors
Where are they located?	Aorta, carotid artery (located in the neck), and the medulla oblongata	Aorta, carotid artery, and vena cava
What do they detect?	Changes in pH	Changes in blood pressure
What responses are produced?	High carbon dioxide concentration in the blood results in low pH, which triggers an increase in heart rate. Low carbon dioxide concentration means a higher pH, which causes a decrease in heart rate.	Heart rate is increased when blood pressure is too low. Heart rate is decreased when blood pressure is higher than normal.

Role of adrenaline

Adrenaline and noradrenaline **increase heart rate** by binding to cardiac cells and increasing the frequency of impulses generated by the SAN.

Synoptic link

You learned about the structure of the heart and the role of the SAN in controlling the cardiac cycle in Topic 8.5, The heart.

Revision tip: Many medullas

A **medulla** is the inner region of a tissue or organ. Other organs, such as the kidney, contain a medulla. Try to refer to the medulla oblongata by its full name to avoid confusion.

Revision tip: Regulation rather than initiation

The cardiac muscle is myogenic, which means its contractions are initiated within the heart tissue rather than by an external stimulus. The accelerator and vagus nerves do not make the heart muscle contract, they regulate the rate at which the SAN fires.

Synoptic link

Topics 14.1, Hormonal communication, and 14.5, Coordinated responses, examine how adrenaline increases heart rate.

Maths skill: Student's *t*-test

A *t*-test is used to assess whether a significant difference exists between two sets of data. Heart rate data lend themselves to this form of analysis. For example, we could compare heart rate before and after exercise, or compare the average resting heart rate of smokers and non-smokers.

You can remind yourself of how to calculate *t* by revisiting Topic 10.6, Representing variation graphically.

Here are some points to remember when using a *t*-test:

- Two data sets are required.

- Mean values are compared.

- The probability value at which biologists tend to assign significance is 0.05. This indicates that there is only a 5% probability that the difference between two data sets is due to chance.

Summary questions

1   Describe the role of receptors in regulating heart rate.   (*3 marks*)

2   Explain why heart rate must be altered in response to increased physical activity.   (*3 marks*)

3   Imagine that the vagus nerve of an organism is cut. Suggest what would happen if the organism's blood pressure increases above its normal value.   (*3 marks*)

1 Which of the following processes is a result / are results of the
secretion of glucagon from the pancreas? *(1 mark)*

    1 Glycogenolysis

    2 Gluconeogenesis

    3 Glycogenesis

    A 1, 2, and 3 are correct

    B Only 1 and 2 are correct

    C Only 2 and 3 are correct

    D Only 1 is correct

2 Which of the following characteristics is generally true of
type 1 diabetes? *(1 mark)*

    A Slow speed of onset.

    B Late onset.

    C Caused by an autoimmune response.

    D Insulin-independent.

3 The set point for blood glucose concentration in humans is
approximately $9 \times 10^{-4}\,\text{g cm}^{-3}$. How would this value be expressed
in $\text{mg}\,100\,\text{cm}^{-3}$? *(1 mark)*

    A 0.9

    B 9

    C 90

    D 900

4 Describe how heart rate is controlled by both nervous and
hormonal mechanisms. *(6 marks)*

5

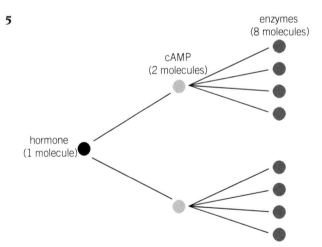

Use the diagram to describe and explain how adrenaline can increase
blood glucose concentration. *(5 marks)*

6 Copy and complete the table to show the differences between type 1 and
type 2 diabetes.

Trait	Type 1	Type 2
Treatment		
Extent of genetic influence		
Is insulin produced?		

(3 marks)

The maintenance and control of an organism's internal conditions within narrow limits is known as **homeostasis**.

Negative feedback

Homeostasis relies on control systems that detect changes in internal conditions and produce the required responses to reverse these changes. The corrective mechanism that allows only small fluctuations around a set point is known as **negative feedback**.

A control system must have the following features:

- A **set point**: this represents the desired value around which the negative feedback mechanism operates. Physiological factors tend to vary over a small range either side of the set point. This represents the **normal range** of that factor.

- **Receptors**: they detect stimuli and deviations from the set point.

- **Controller** (communication pathway): this coordinates the information from the receptors and sends instructions to the effectors. The nervous system and the hormonal system tend to act as controllers.

- **Effectors**: they produce the changes required to return the system to the set point.

- **Feedback loop**: the return to the set point creates a feedback loop.

> **Key term**
>
> **Homeostasis**: The maintenance of stable conditions (within narrow limits) inside the body.

> **Key term**
>
> **Negative feedback**: The mechanism controlling homeostasis; a change in a parameter leads to the reversal of the change.

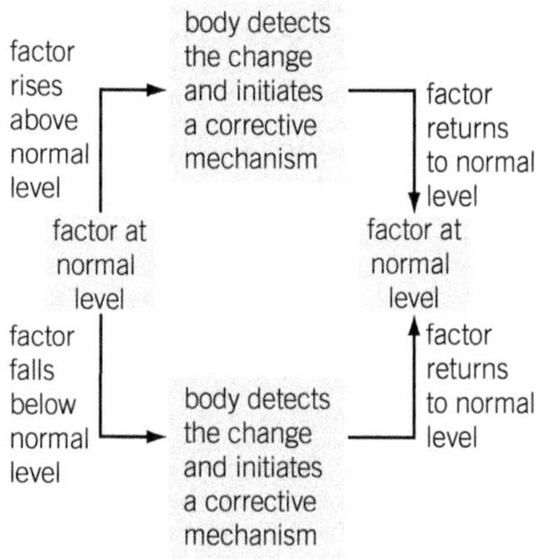

factor rises above normal level → body detects the change and initiates a corrective mechanism → factor returns to normal level

factor at normal level

factor at normal level

factor falls below normal level → body detects the change and initiates a corrective mechanism → factor returns to normal level

▲ **Figure 1** *Negative feedback*

 Worked example: Applying terms to an example of homeostasis

Apply the terms 'set point', 'normal range', 'receptors', 'controller', and 'effectors' to the homeostatic control of water potential in the blood (see Topic 15.5, The structure and function of the mammalian kidney).

Set point = 285 mOsm kg^{-1}

Normal range = 275–295 mOsm kg^{-1}

Receptors = osmoreceptors (in the hypothalamus)

Controller = hypothalamus (secreting ADH)

Effectors = cells of the collecting duct (in the kidneys)

Positive feedback

Negative feedback reverses a change in conditions, whereas positive feedback increases the original change. Negative feedback **inhibits** the original stimulus, but positive feedback **enhances** the original stimulus.

Positive feedback is not involved in homeostasis. Examples include:

- the attraction of platelets to a site of blood clotting
- the generation of action potentials in neurones (see Topic 13.4, Nervous transmission)
- childbirth (the hormone oxytocin is released when the baby's head pushes against the cervix; oxytocin stimulates uterine contractions, which causes more of the hormone to be produced).

Summary questions

1 In the context of a control system, state what is meant by the terms **a** set point **b** normal range. *(2 marks)*

2 Describe the positive feedback mechanism that controls uterine contractions during labour, and suggest how positive feedback is beneficial during this process. *(4 marks)*

3 Explain why homeostasis relies on negative feedback and not positive feedback. *(3 marks)*

Synoptic link

The principles of homeostasis apply to several topics in the course, such as the regulation of blood glucose (Topic 14.3), heart rate (Topic 14.6), thermoregulation (Topic 15.2), and osmoregulation (Topic 15.5).

Revision tip: *Almost constant*

Homeostasis does not maintain *constant* values. Continuous fluctuations in physiological conditions occur. The important point is that homeostasis maintains these fluctuations within a narrow range.

15.2 Thermoregulation in ectotherms
15.3 Thermoregulation in endotherms

Specification reference: 5.1.1 (d)

You learned about the general principles of homeostasis in the previous topic. One example of homeostasis is **thermoregulation** – the control of body temperature.

In ectotherms

Environmental temperature rather than metabolism dictates body temperature in **ectotherms** (animals such as fish, amphibians, and reptiles). Ectotherms show behavioural responses to regulate their internal temperature.

▼ **Table 1** *Ectotherm behaviours*

Behaviour when too cold	Behaviour when too hot
Basking (exposing their bodies to the Sun)	Finding shade or burrowing
Change body shape (e.g. increase surface area to gain heat in hot weather; decrease surface area to retain heat in cold weather)	
Pressing body against warm ground (to gain heat through conduction)	Pressing body against cool stones (to lose heat through conduction)

In endotherms

Endotherms (mammals and birds) use internal mechanisms to control body temperature.

The homeostatic control of body temperature relies on the following components:

Receptors: peripheral temperature receptors (in skin), temperature-sensitive neurones in the hypothalamus (to monitor core temperature).

Controller: hypothalamus

Effectors: e.g. sweat glands, erector muscles controlling hair follicles, skeletal muscle, sphincter muscles controlling vasodilation and vasoconstriction in arterioles.

▼ **Table 2** *Endotherm responses to internal temperature changes – these are negative feedback responses controlled by the hypothalamus*

To warm up	Explanation	To cool down	Explanation
Less sweat	Less heat lost through evaporation of sweat	More sweat	Sweat evaporation requires heat from the blood, producing a cooling effect
Hairs raised	A layer of insulating air is trapped	Hairs lie flat on the skin	More heat can be lost through radiation
Vasoconstriction	Less blood flows through capillaries near the skin surface; less heat is radiated from the body	Vasodilation	More blood flows through capillaries near the skin surface; more heat is radiated from the body
High metabolic rate in liver cells	Respiration generates more heat	Low metabolic rate in liver cells	Respiration generates less heat
Skeletal muscles contract spontaneously (shivering)	Heat generated from respiration	No spontaneous contractions	No additional heat generated from respiration in muscles

Key term

Ectotherm: An animal that depends on external sources of heat to determine body temperature.

Key term

Endotherm: An animal that can use internal sources of heat to control body temperature.

Revision tip: The benefit of endothermy

The maintenance of a stable internal temperature, irrespective of changes in surrounding environmental temperature, has enabled endotherms to occupy habitats with a wide range of ambient temperatures.

Revision tip: Ectotherm physiological responses and endotherm behaviours

In general, ectotherms use behaviours to warm up or cool down, whereas endotherms rely on physiology to a greater extent. However, some ectotherms have evolved physiological adaptations (e.g. dark pigments to absorb radiation, and the ability to alter metabolic rate), and endotherms tend to show the same range of behavioural responses exhibited by ectotherms.

Summary questions

1 Name four structures located in the skin that are used by endotherms in thermoregulation. (*4 marks*)

2 Discuss the advantages and disadvantages of ectothermy. (*4 marks*)

3 Suggest why the food intake of an endothermic species is likely to increase during the winter compared to the summer months. (*3 marks*)

15.4 Excretion, homeostasis, and the liver

Specification reference: 5.1.2(a) and (b)(i–ii)

Excretion is the removal of waste products from the body. The liver is an organ that plays an important role in excretion, as well as having a range of other functions.

Excretion

Examples of substances that need to be excreted include:

- Carbon dioxide (see Topics 7.1 to 7.4).
- Urea ($CO(NH_2)_2$), which is produced in the liver from ammonia and carbon dioxide).
- Bile pigments (which form when haemoglobin is broken down by **Kupffer cells** – specialised macrophages in the liver).

The liver

Structure

The liver is divided into units called **lobules**. Each lobule contains liver cells (**hepatocytes**), which border two types of channel: **sinusoids** (vessels in which blood from the hepatic artery and hepatic portal vein mix) and **bile canaliculi** (which drain bile into the gall bladder).

Functions of the liver

Function	Description
Deamination	An amino acid can be converted to ammonia (NH_3) and a keto acid. Ammonia is then converted to urea in the **ornithine cycle** ($2NH_3 + CO_2 \rightarrow CO(NH_2)_2 + H_2O$).
Transamination	The conversion of one amino acid to another.
Glycogen storage	Glucose is converted to glycogen in hepatocytes; glycogen stores are hydrolysed to glucose when required (see Topic 14.3).
Detoxification	e.g. hydrogen peroxide (a metabolic waste product) is converted to water and oxygen by catalase. e.g. ethanol is converted to ethanal by alcohol dehydrogenase.

> **Common misconception: Excretion, egestion, secretion**
>
> **Excretion** is the removal of metabolic waste from the body. Do not confuse this with **egestion** (i.e. defecation – the removal of undigested material from the body) or **secretion** (i.e. the release of a substance for a particular function – e.g. sweat secretion).

> **Summary questions**
>
> 1 State two examples of substances that the liver removes in its role as an organ of detoxification. *(2 marks)*
>
> 2 Explain the importance of transamination. *(2 marks)*
>
> 3 Describe the role of enzymes in aspects of liver function. *(5 marks)*

> **Key term**
>
> **Excretion:** The removal of metabolic waste products from the body.

bile canaliculus – the hepatocytes secrete bile into these canaliculi and from there it drains into the gall bladder to be stored

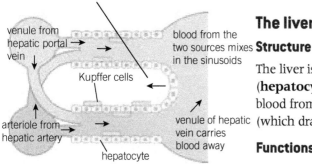

venule from hepatic portal vein

blood from the two sources mixes in the sinusoids

Kupffer cells

arteriole from hepatic artery

venule of hepatic vein carries blood away

hepatocyte

▲ **Figure 1** *The structure of a liver lobule*

> **Revision tip: Blood ties**
>
> The liver is supplied by blood from two vessels:
>
> 1 The **hepatic artery** supplies oxygenated blood from the heart.
>
> 2 The **hepatic portal vein** supplies blood carrying glucose, amino acids, and fats from the small intestine.
>
> However, the **hepatic vein** transports deoxygenated blood *from the liver*.

> **Synoptic link**
>
> You may be required to examine and draw liver tissue under a microscope. Remind yourself of some basic microscopy skills by reading Topic 2.1, Microscopy.

15.5 The structure and function of the mammalian kidney

Specification reference: 5.1.2(c)i–iii, (d), (e), and (f)

Mammalian kidneys perform two homeostatic functions: **excretion** (see Topic 15.4, Excretion, homeostasis, and the liver) and **osmoregulation** (the control of water and ion concentrations in an organism).

Kidney structure

The two kidneys in vertebrates are supplied with blood by the **renal artery**, and a **renal vein** drains each organ. **Urine** produced by the kidneys passes through the **ureter** to the **bladder**.

Each kidney has three distinct regions:

- The **cortex** (the outer region).
- The **medulla** (the inner region).
- The **renal pelvis** (the most central region, which leads to the ureter).

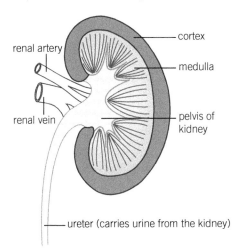

▲ **Figure 1** *The gross structure of a kidney*

The functional unit of the kidney is called a **nephron**; each kidney contains approximately one million nephrons.

Synoptic link

You may be required to examine and draw kidney tissue under a microscope. Remind yourself of some basic microscopy skills by reading Topic 2.1, Microscopy.

Revision tip: Three filters

Three different filters prevent cells and large molecules from leaving the blood during ultrafiltration:

Narrow gaps (**fenestrations**) between endothelium cells in capillaries.

Podocytes (epithelial cells of the capsule that have finger-like projections, which form filtration slits).

Basement membrane (a mesh of collagen and glycoprotein around the glomerulus).

Functions of regions of the nephron

Region	Function	How does it work?
Bowman's capsule	Ultrafiltration	Water and other small molecules are forced from the blood (a network of capillaries called the **glomerulus**) into the Bowman's capsule. Large molecules and blood cells remain in the glomerulus. High **hydrostatic pressure** is created in the glomerular capillaries by the **afferent** (incoming) arteriole being wider than the **efferent** (outgoing) arteriole. The fluid that is filtered into the capsule is called glomerular **filtrate**.
Proximal convoluted tubule (PCT)	Selective reabsorption	All glucose, amino acids, vitamins, and hormones are reabsorbed into the blood. 85% of water and NaCl is reabsorbed. The cells of the PCT wall are specialised for reabsorption in the following ways: • **Microvilli** to increase surface area for reabsorption • Plasma membranes have many **pumps** and **transporter proteins** for active transport and facilitated diffusion • Many **mitochondria** to produce ATP for active transport. Toxic compounds (e.g. urea), excess water, and some ions remain in the filtrate.
Loop of Henle	Establishing a water potential gradient	The filtrate and blood are **isotonic** at the beginning of the loop of Henle (i.e. no water potential gradient exists). By the end of the loop of Henle, the water potential in the filtrate is higher than in the medulla. This enables additional water to be reabsorbed from the collecting duct down a **water potential gradient**, if necessary. The **descending limb** is permeable to water. However, the **ascending limb** is impermeable to water, but Na^+ and Cl^- ions are actively transported from the filtrate into the medulla, which raises the water potential of the filtrate.
Distal convoluted tubule	Additional ion reabsorption	Additional active transport and reabsorption of ions can occur here.
Collecting duct	Determination of urine concentration and volume	Water can be reabsorbed, depending on the body's needs. The permeability of the collecting duct (and therefore the volume of water that is reabsorbed) is controlled by ADH (see below).

Osmoregulation

Water potential is controlled through a negative feedback mechanism. A decrease in blood water potential below the set point causes the following response:

• **Osmoreceptors** in the **hypothalamus** detect the low water potential.
• Neurosecretory cells (specialised nerve cells) in the hypothalamus are stimulated to release **ADH** from the **posterior pituitary gland**.

- ADH travels through the blood to the kidneys.
- ADH binds to receptors on the cell surface membrane of cells in the **collecting duct wall**.
- The concentration of cyclic AMP (**cAMP**) in these cells is increased.
- cAMP acts as a second messenger and causes **aquaporins** to be inserted into the membrane of cells in the collecting duct wall.
- Aquaporins are membrane-spanning channel proteins that increase the permeability of the collecting duct wall. They allow water to diffuse through but prevent the passage of ions.
- More water is reabsorbed from the collecting duct by osmosis.
- Urine with a high solute concentration is produced.
- The water potential of the blood is increased.

The same system responds if blood water potential rises above the set point. Less ADH is produced, and less water is reabsorbed into the blood from the collecting duct. A large volume of dilute urine is produced.

Revision tip: Urea vs. urine

Make sure you know the difference between urea and urine. Urea is the waste molecule $CO(NH_2)_2$, whereas urine is the fluid produced by the kidneys that contains dissolved urea.

Go further: Diuretics – treating, cheating, and energy-depleting?

A **diuretic** is a substance that increases the volume of urine produced by an individual. There are many examples, which produce their effects through a variety of different mechanisms.

Diuretics have many medical uses. High blood pressure (hypertension) can be treated with thiazide diuretics. Thiazides target cells in the distal convoluted tubule wall. They inhibit sodium–chloride ion cotransport, which leads to more water being excreted in the urine.

Some athletes use diuretics to cheat by invalidating their drug tests. By raising their urine volume, athletes can dilute performance-enhancing drugs and their metabolites. The diuretics act as masking agents.

Alcohol acts as a diuretic by inhibiting ADH production. Hangovers are not only a result of alcohol toxicity; they are also caused by the effect on brain cells of blood with a high solute concentration.

1 Suggest why thiazide diuretics result in a higher volume of urine being excreted.

2 Explain how drinking plenty of water after heavy alcohol consumption could limit the effects of alcohol on brain cells.

Uses of excretory products

The excretory products in urine can be analysed to make medical diagnoses.

Excretory product	What is diagnosed?	Details
Glucose	Diabetes	Glucose should be reabsorbed in the PCT of nephrons. The presence of glucose may indicate diabetes.
Human chorionic gonadotrophin (hCG)	Pregnancy	**Monoclonal antibodies** detect the presence of hCG (which is produced during pregnancy) in urine.
Anabolic steroids (and other drugs)	The use of banned drugs in sport	Urine samples are tested using gas chromatography and mass spectrometry.

Kidney failure

Kidney failure is either acute or chronic. Kidneys can fail for the following reasons: bacterial infection, kidney stones, uncontrolled diabetes (type 1 or 2), hypertension, and certain inherited diseases.

Treatments

Treatment	Description	Advantages	Disadvantages
Transplant	Surgery to replace a failed kidney with a donor's kidney	If successful, removes the need for dialysis treatment	Possible transplant rejection (therefore patient requires immunosuppressant drugs)
Haemodialysis	A patient's blood is filtered through a dialysis machine	Daily dialysis is not required	Patients must spend hours in hospitals each week
Peritoneal dialysis	A patient's abdominal membrane (peritoneum) is use as a dialysis membrane	It can be done at home, and patients are able to walk and work during dialysis	Must be performed daily and requires an initial implantation of a tube

Revision tip: Acute vs. chronic

Remember that an **acute** condition has a quick onset but often a fast recovery time. A **chronic** condition tends to develop more slowly but usually lasts a long time.

Summary questions

1 Outline the likely differences in composition between blood in the renal artery and blood in the renal vein. *(2 marks)*

2 The cells lining the proximal convoluted tubule use endocytosis and exocytosis, in addition to active transport and facilitated diffusion, to move molecules across their membranes. Suggest why endocytosis and exocytosis are used. *(3 marks)*

3 Evaluate the costs and benefits of dialysis and transplant surgery as treatments for kidney failure. *(6 marks)*

1 Which of the following statements is an example / are examples
   of positive feedback? *(1 mark)*
   1 The mechanism by which platelets are attracted to a site of blood clotting.
   2 The interaction between oxytocin and contraction strength during childbirth.
   3 The development of hypothermia.
   A 1, 2, and 3 are correct
   B Only 1 and 2 are correct
   C Only 2 and 3 are correct
   D Only 1 is correct

2 Which of the following is a response to the body's temperature
   rising above its set point? *(1 mark)*
   A Vasoconstriction.
   B Hairs lie flat.
   C Increased muscle contraction within some tissues.
   D Decreased sweat production.

3 Which of the following is an adaptation / are adaptations in a
   nephron to enable ultrafiltration? *(1 mark)*
   1 Cells in capillary walls are separated by narrow gaps.
   2 The endothelium cells of the capillaries have finger-like projections called podocytes.
   3 Basement membranes allow large proteins to pass through into the nephron filtrate.
   A 1, 2, and 3 are correct
   B Only 1 and 2 are correct
   C Only 2 and 3 are correct
   D Only 1 is correct

4 In which region of a nephron does the majority of selective
   reabsorption occur? *(1 mark)*
   A Proximal convoluted tubule.   C Distal convoluted tubule.
   B Loop of Henle.   D Collecting duct.

5 Which of the following is / are true of antidiuretic hormone? *(1 mark)*
   1 It is released from the posterior pituitary gland.
   2 It binds to cells in the collecting duct wall.
   3 It results in aquaporins being inserted into the membrane of cells in the collecting duct wall.
   A 1, 2, and 3 are correct   C Only 2 and 3 are correct
   B Only 1 and 2 are correct   D Only 1 is correct

6 Which of the following statements is true of peritoneal dialysis? *(1 mark)*
   A It is usually performed three times per week.
   B A patient's blood is passed through a dialysis machine and returned to their body.
   C It must be performed every day.
   D It must be carried out by medical professionals.

16.1 Plant hormones and growth in plants

Specification reference: 5.1.5(b), (c), (d) and (e)

Synoptic link

You learned about mammalian hormones in Chapter 14, Hormonal communication.

Like animals, plants produce hormones. These chemicals enable cell signalling and communication between different parts of a plant. Hormones work in concert to create coordinated responses to environmental conditions.

Roles of hormones

Auxins and **gibberellins** are two families of hormones responsible for a range of functions.

Hormone	Role	How does it operate?	Experimental evidence
Auxins	Cell elongation	Increases cell wall stretchiness, enabling water absorption and cell expansion	Auxin application causes pH to decrease. Enzymes that increase cell wall flexibility work best in acidic conditions.
	Apical dominance	Promotes growth of the main shoot and inhibits lateral shoots	Lateral shoots grow faster when the apex (i.e. tip) of the main shoot (the site of auxin production) is removed.
	Root growth	Low auxin concentrations promote root elongation	Application of small quantities of auxin stimulates root growth; high concentrations inhibit growth.
Gibberellins	Stem elongation	Stimulates cell elongation and division	Gibberellin concentrations and plant height show a positive correlation.
	Seed germination	Activates genes for amylase and protease enzymes, which break down food stores in seeds	Mutant seeds that lack gibberellin genes and seeds exposed to gibberellin inhibitors do not germinate.

Another hormone, **ethene**, promotes leaf fall (abscission) and fruit ripening. **Abscisic acid** (ABA) works in opposition to gibberellins to stop seed germination.

Revision tip: Concentrate on the units

The amount of a substance is measured in moles, and its concentration tends to be measured in $mol\,dm^{-3}$ (the number of moles in a cubic decimetre of solution). A decimetre is a tenth of a metre (0.1 m), whereas a centimetre is a hundredth of a metre (0.01 m). When dealing with volumes, this means $1\,dm^3 = 1000\,cm^3$. The amount of substance (in moles) can be found by calculating volume × concentration.

Maths skill: Serial dilutions

The effect of plant hormones on growth can be investigated by creating different concentrations of hormone solution. This is achieved using serial dilutions, as illustrated in the following example:

- Solution X has a volume of $10\,cm^3$ and a concentration of $1\,mol\,dm^{-3}$
- $1\,cm^3$ of X is added to $9\,cm^3$ of distilled water. This produces a $0.1\,mol\,dm^{-3}$ solution (Y)
- $1\,cm^3$ of Y is added to $9\,cm^3$ of distilled water. This produces a $0.01\,mol\,dm^{-3}$ solution

Summary questions

1 State three roles of auxins. *(3 marks)*

2 Using your knowledge from Chapter 14, suggest two similarities and two differences between the ways in which plant and animal hormones operate. *(4 marks)*

3 A gibberellin solution with a concentration of $0.5\,mol\,dm^{-3}$ and a volume of $50\,cm^3$ was diluted. $1\,cm^3$ of the solution was added to $9\,cm^3$ of water. From this second solution, $1\,cm^3$ was removed and mixed with $9\,cm^3$ of water. Calculate the concentration and number of moles in the final solution. *(2 marks)*

16.2 Plant responses to abiotic stress
16.3 Plant responses to herbivory

Specification reference: 5.1.5(a)(i) and (b)

Plants experience abiotic stresses. They must respond to environmental fluctuations, such as changes in day length and temperature. They also experience biotic stress – consumption by other organisms is a threat that looms ever-present for many plant species. Plants have evolved responses to cope with the stresses of environmental change and herbivory.

Plant responses

	Factor	Plant response	Mechanism
Abiotic	Day length change	Leaf loss (abscission)	Reduced light exposure → reduced auxin → increased ethene sensitivity → increased cellulase production → cell walls digested at abscission zone.
		Flowering (either when days become shorter or longer)	The relative proportion of two phytochrome photoreceptors (P_{fr} and P_r) dictates when flowering begins. P_{fr} accumulates in the day and is converted back to P_r at night.
	Cold weather	Solutes produced as antifreeze	Reduced temperatures and day length switch certain genes on or off (i.e. epigenetic control of transcription).
	Water shortage	Stomatal closure	The release of ABA causes stomata to close.
Biotic	Herbivory	Physical defences	E.g. the evolution of thorns, barbs, spikes, and inedible tissue.
		Chemical defences	E.g. toxic compounds such as tannins, alkaloids, and terpenoids. Pheromones warn other plants (or regions of the same plant) of an attack.

Go further: Flowering – the long and the short of it

The balance between P_{fr} and P_r proteins is thought to determine the onset of flowering in plants. P_{fr} is produced during the day. P_r is produced during the night. Some species (long-day plants) flower when day length increases. Other species (short-day plants) flower as day length decreases.

1   Suggest how the relative proportions of P_r and P_{fr} change to initiate flowering in short-day and long-day plants.

Synoptic link

In Topic 9.5, Plant adaptations to water availability, and 12.4, Plant defences against pathogens, you learned about the evolution of other plant adaptations.

Summary questions

1   Describe the defences plants have evolved against herbivory. (*4 marks*)

2   Explain how leaf fall is initiated in some plants as day length decreases. (*4 marks*)

3   Using information from this topic and Topic 12.4, Plant defences against pathogens, suggest how a plant is able to prepare for a pathogenic attack when a neighbouring plant becomes infected. (*2 marks*)

Key term

Tropism: Growth or movement in response to an environmental stimulus.

A **tropism** is the growth of a plant in a particular direction. Plants grow towards sources of light, which is known as **phototropism**. Growth in response to gravity is called **geotropism**.

Types of tropism

Tropisms can be **positive** or **negative**. Shoots demonstrate negative geotropism (growing against gravity) and positive phototropism (growing towards light), whereas roots show positive geotropism and negative phototropism.

Question and model answer: Phototropism

Q. Describe the mechanism that controls phototropism in plants, including the role of auxins.

A.

- The shoot **apex** produces **auxin**, which diffuses (or is actively transported) down the shoot.

- Auxin concentration will be **greater on the shaded side** of the shoot/the side receiving less light.

- Auxin causes shaded **cells to elongate**/expand because their **cell walls are loosened**/stretched.

- The cell walls are loosened because **H⁺ ions** are pumped into the walls, **lowering pH**, which increases the activity of **expansins** (proteins that mediate the loosening of cell walls).

- Cells elongate more on the shaded side than the brighter side of the shoot.

- The shoot therefore bends towards the light.

Common misconception: Tropism not trophism

Do not confuse 'tropism' with 'trophism', which is an alternative word for 'nutrition' (see Topic 23.2, Biomass transfer through an ecosystem).

Practical skill: Investigating tropisms

Phototropism can be investigated using the following approaches:

- Vary **light intensity** (e.g. seedlings can be grown in darkness, all-round light, or unilateral light (i.e. light from one side of the plant)).

- Grow seedlings in unilateral light and vary **light wavelengths**, using filters.

- Vary **auxin concentration** (e.g. by removing the shoot apex or introducing additional auxin to one side of the plant).

Geotropism can be demonstrated by rotating a growing plant in a **clinostat**. The shoot and root should grow straight because the gravitational pull is applied evenly to the plant.

Summary questions

1 Describe how geotropism can be demonstrated using a clinostat.
(2 marks)

2 Describe the role of active transport in positive phototropism. *(3 marks)*

3 Explain the importance of positive and negative phototropism to plant survival.
(4 marks)

16.5 The commercial use of plant hormones

Specification reference: 5.1.5(f)

The production of hormones in plants is often dictated by environmental conditions. For example, ABA production is increased when water is in short supply. However, scientists are now able to harness their knowledge of plant biology and manipulate hormone concentrations for commercial benefit.

How can hormones be used?

Hormone	Commercial uses
Ethene	Speeds up **ripening** Promotes **fruit dropping** (e.g. in walnuts and cherries)
Auxins	Slows down **leaf fall** and **fruit dropping** (although high concentrations actually promote fruit dropping) Encourages **root growth** (in cuttings) Synthetic auxins can act as **weedkillers**
Cytokinins	**Prevents senescence** (ageing)
Gibberellins	**Delays senescence** (in citrus fruit) Speeds up barley **seed germination** during alcohol brewing

Synoptic link

In later topics (e.g. Topic 22.1, Natural cloning in plants, and Topic 22.2, Artificial cloning in plants) you will examine other uses of plant hormones.

Summary questions

1   State three commercial uses of auxin.   *(3 marks)*

2   Solutions of 2-chloroethylphosphonic acid are often applied to plants instead of ethene, which is a gas. Suggest why.   *(2 marks)*

3   What can you conclude from Figure 1 about the effect of ethene on fruit ripening? Explain your answer.   *(2 marks)*

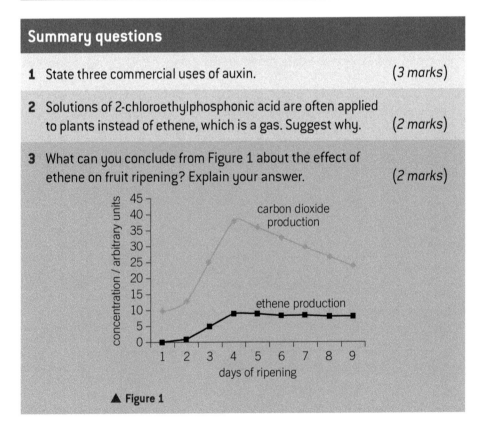

▲ Figure 1

Chapter 16 Practice questions

1 Which of the following scenarios is/are likely to result in flowering in short-day plants? *(1 mark)*

   1 An increase in night length beyond a critical period.

   2 An increase in day length beyond a critical period.

   3 An increase in P_{fr} concentration within the plant.

   A 1, 2, and 3 are correct

   B Only 1 and 2 are correct

   C Only 2 and 3 are correct

   D Only 1 is correct

2 Which of the following statements is/are true of gibberellins? *(1 mark)*

   1 They promote seed germination.

   2 They promote stem elongation.

   3 Their action is inhibited by ABA.

   A 1, 2, and 3 are correct

   B Only 1 and 2 are correct

   C Only 2 and 3 are correct

   D Only 1 is correct

3 The graph below shows the effect of gibberellin exposure on amylase production in isolated tissue from barley seeds.

   Describe and explain the data shown in the graph. *(5 marks)*

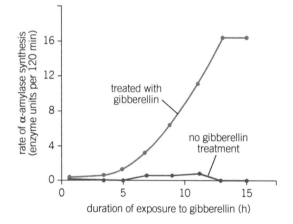

4 Suggest the significance of the data in the graph below to scientists carrying out tissue culture. *(2 marks)*

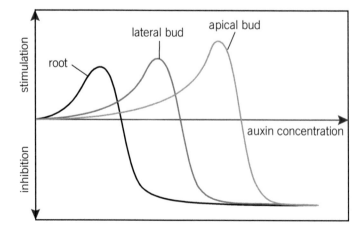

17.1 Energy cycles

Specification reference: 5.2.1(a) and 5.2.2(a)

Energy is required for many biological processes, including active transport, movement, and anabolic reactions. Energy enters ecosystems through photosynthesis, in which plants trap solar energy as chemical energy. Cellular respiration converts the chemical energy to a useable form: ATP. Here you will examine the relationship between photosynthesis and respiration.

A comparison of respiration and photosynthesis

Although the two processes involve many different reactions, the overall equations for photosynthesis and respiration mirror each other. Photosynthesis uses light energy to convert inorganic molecules (carbon dioxide and water) to organic molecules (e.g. glucose) and oxygen. Respiration converts glucose back to carbon dioxide and water, which releases the energy that is used to generate ATP.

	Photosynthesis	Respiration
Purpose	Conversion of light energy to chemical energy in organic molecules $6CO_2 + 6H_2O \rightarrow C_6H_{12}O_6 + 6O_2$	Conversion of chemical energy in organic molecules to chemical energy in ATP $C_6H_{12}O_6 + 6O_2 \rightarrow 6CO_2 + 6H_2O$
Reactants	Carbon dioxide and water	Glucose and oxygen
Products	Glucose and oxygen	Carbon dioxide and water
Type of reaction	Endothermic (overall, energy is taken in)	Exothermic (overall, energy is released)
ATP production	Produced in the light-dependent stage and used in the light-independent stage	An end product
Use of coenzymes	NADP carries H atoms between the two stages of photosynthesis	NAD and FAD carry H atoms to the electron transport chain

Summary questions

1 Complete the following passage by choosing the most appropriate words to place in the gaps.
Photosynthesis is .... (i.e. energy is taken in, overall). It converts light energy to chemical energy, which is required by organisms for processes such as .... transport and anabolic reactions. One product of photosynthesis is glucose, which is broken down during cellular respiration to produce ....; this molecule is the currency of chemical energy used by all cells. *(3 marks)*

2 State three similarities between the processes of photosynthesis and respiration. *(3 marks)*

3 Describe the energy conversions and transfers that occur between light being absorbed by chlorophyll in plant leaves and heat being radiated from organisms. *(3 marks)*

17.2 ATP synthesis

Specification reference: 5.2.2(h)

Synoptic link

Topic 3.11 , ATP, outlined the structure of ATP.

ATP is the currency of chemical energy used by all species. Cellular respiration generates ATP as an end product. However, ATP is also produced by the light-dependent stage of photosynthesis. In both cases, ATP production relies on electron transport chains, chemiosmosis, and ATP synthase.

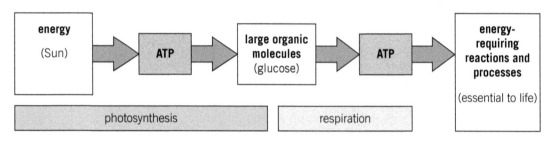

▲ **Figure 1** An overview of ATP's role in photosynthesis and respiration

Key term

Chemiosmosis: The movement of H⁺ ions down an electrochemical gradient, which drives the production of ATP in both respiration and photosynthesis.

Summary questions

1 State two differences between chemiosmosis in respiration and photosynthesis. (*2 marks*)

2 Describe how a proton gradient is established across a thylakoid membrane during photosynthesis. (*3 marks*)

3 Suggest how the following research findings provide evidence for the chemiosmotic theory.
 a The pH of the mitochondrial intermembrane space is lower than that of the matrix. (*2 marks*)
 b The potential difference across the inner membrane is −200 mV. The matrix is more negative than the intermembrane space. (*1 mark*)

Chemiosmosis

In both photosynthesis and respiration, chemiosmosis involves the following steps:

1 Electrons are raised to a higher energy level (i.e. **excited electrons**)

2 The high-energy electrons pass along an **electron transport chain** (i.e. a series of electron carriers)

3 Energy is released as the electrons are passed to lower energy levels

4 The energy is used to pump H⁺ ions (protons) across a membrane

5 A **proton gradient** is established across the membrane

6 Protons move down the concentration gradient through channel proteins linked to **ATP synthase**

7 The flow of protons provides kinetic energy to enable **ATP synthesis** by ATP synthase.

Comparison of chemiosmosis in respiration and photosynthesis

The principles of chemiosmosis are the same in both respiration and photosynthesis, but several differences exist.

▼ **Table 1** A comparison of chemiosmosis in photosynthesis and respiration

	Photosynthesis	Respiration
Where do the high energy electrons come from?	Light is absorbed by chlorophyll	Electrons are released from chemical bonds in glucose
Where is the location of the electron transport chain?	Thylakoid membranes (in chloroplasts)	Inner mitochondrial membranes
What is the name given to ATP production?	Photophosphorylation	Oxidative phosphorylation

17.3 Photosynthesis

Specification reference: 5.2.1(b), (c)i-ii, (d), (e), and (f)

Photosynthesis comprises two stages, both of which occur in chloroplasts. The light-dependent stage relies on a range of pigments to absorb sunlight and convert it to chemical energy. The light-independent stage fixes carbon dioxide and transfers the chemical energy into organic molecules such as glucose.

Chloroplast structure

Chloroplasts consist of two distinct regions:

- **Grana** (singular: granum): flattened membrane compartments (thylakoids), which are the sites of the light-dependent stage of photosynthesis.
- **Stroma**: A fluid-filled matrix, which is the site of the light-independent stage of photosynthesis.

Synoptic link

You learned about the structure of chloroplasts in Topic 2.5, The ultrastructure of plant cells.

▼ **Table 1** *The adaptations of chloroplasts*

Adaptation	Purpose
Thylakoid membranes are stacked	Large surface area over which light-dependent reactions can occur
Photosynthetic pigments are organised into photosystems	The efficiency of light absorption is maximised
Grana are surrounded by the stroma	Products of the light-dependent reactions (reduced NADP and ATP) can pass directly to the enzymes catalysing the light-independent reactions
Chloroplasts contain their own DNA and ribosomes	Photosynthetic proteins can be produced inside chloroplasts rather than being imported
The inner chloroplast membrane is embedded with transport proteins and is less permeable than the outer membrane	Control over which substances that can enter the stroma from the cell cytoplasm

Photosynthetic pigments

- **Photosystems** are light-harvesting complexes of pigments found in thylakoid membranes.
- **Accessory pigments** (e.g. chlorophyll *b*, carotenoids, and xanthophylls) absorb photons of light and funnel this energy to a **reaction centre** at the heart of the photosystem.
- **Chlorophyll *a*** is located in the reaction centre.
- **Electrons** from chlorophyll *a* are excited and passed to electron acceptors at the beginning of the electron transport chain.

Practical skill: Thin layer chromatography

Photosynthetic pigments can be separated and identified using a technique called **chromatography**.

How are pigments separated? The pigment molecules interact with the stationary phase to different extents and therefore move at different rates.

What is the stationary phase? Either silica gel on glass, or paper.

What is the mobile phase? A solvent in which the pigment molecules dissolve.

Pigments can be identified and compared by calculating R_f values:

$$R_f = \frac{\text{distance travelled by pigment}}{\text{distance travelled by solvent}}$$

▼ **Table 2** *An example of R_f calculations*

Pigment	Distance travelled (cm)	Solvent distance (cm)	R_f
carotene	4.75	5.00	0.95
chlorophyll a	3.25	5.00	0.65
chlorophyll b	2.25	5.00	0.45

Synoptic link

You encountered chromatography in the context of separating amino acids in Topic 3.6, Structure of proteins.

Synoptic link

NADP is a coenzyme. You learned about coenzymes in Topic 4.4, Cofactors, coenzymes, and prosthetic groups.

Revision tip: A numbers game

It might seem odd that photosystem II appears before photosystem I in the light-dependent stage. The names are a result of photosystem I being identified first.

Revision tip: Cyclic phosphorylation

Cyclic phosphorylation produces ATP but not reduced NADP. Only photosystem I (PSI) is involved. Electrons are excited from PSI, enabling ATP to be generated from an electron transport chain. Instead of being used to generate reduced NADP, the electrons return to PSI.

Revision tip: P for photosynthesis

NADP is the coenzyme used in photosynthesis, whereas NAD is used in respiration. You can remember *P* for photosynthesis. It is acceptable to write reduced NADP as NADPH.

The light-dependent stage

The purpose of the light-dependent stage of photosynthesis is to generate two products, which then feed into the **Calvin cycle**.

- **Light energy** is absorbed and enables **ATP production**.
- **Hydrogen from water** is used to **reduce NADP**.

Non-cyclic phosphorylation in the light-dependent stage involves the following steps:

1 Electrons from the reaction centre in photosystem II are excited
2 The excited electrons are passed along an electron transport chain (and ATP is produced)
3 The electrons from photolysis (the splitting of water using light energy) replace those lost from photosystem II
4 Electrons from the reaction centre in photosystem I are excited
5 More ATP is produced via a second electron transport chain
6 The electrons that left photosystem II replace those lost from photosystem I
7 The electrons from photosystem I and H⁺ ions from the photolysis of water combine to reduce NADP.

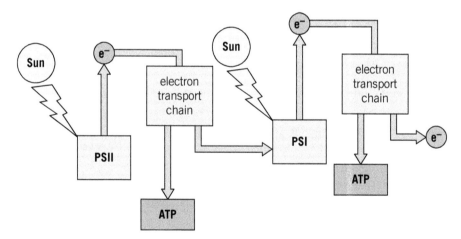

▲ **Figure 1** *Non-cyclic phosphorylation*

The light-independent stage (the Calvin cycle)

Hydrogen (from reduced NADP) and carbon dioxide are used to synthesise glucose and other organic molecules. The energy required for these reactions is supplied by ATP.

The Calvin cycle involves the following steps:

1 The enzyme RuBisCo catalyses the reaction between RuBP (a 5-carbon sugar) and CO_2 to produce two molecules of three-carbon GP
2 ATP and NADPH are used to reduce GP to another three-carbon molecule, TP
3 The majority of TP is used to regenerate RuBP, which continues the cycle. More ATP is required for this reaction.

Uses of triose phosphate

Other than being used to regenerate RuBP, triose phosphate (TP) can be converted into a wide range of products:

- Six-carbon sugars (e.g. glucose and fructose)
- Fructose and glucose can react to produce the disaccharide sucrose
- Glucose molecules can react together to form polysaccharides such as amylose, amylopectin, and cellulose
- A single TP molecule can be converted to glycerol
- TP can act as the starting point for the synthesis of amino acids.

➕ Go further: A visit to the GP

Figure 2 shows some of the molecules that can be formed from GP. We can see that only small changes occur during these reactions. When GP is converted to the intermediate 1,3-diphosphoglycerate, the OH (hydroxyl) group is replaced by a phosphate group. When 1,3-diphosphoglycerate is converted to TP, this phosphate is substituted for a hydrogen atom. Figure 2 also shows the conversion of GP into another important biological molecule, serine.

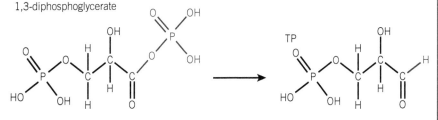

▲ **Figure 2** *An outline of molecules that can be produced from GP*

1 Which molecules are required for the conversion of (a) GP to 1,3-diphosphoglycerate (b) 1,3-diphosphoglycerate to TP?

2 Serine is an amino acid. Draw the R group.

3 Suggest a mineral ion that would be required during the conversion of GP to serine.

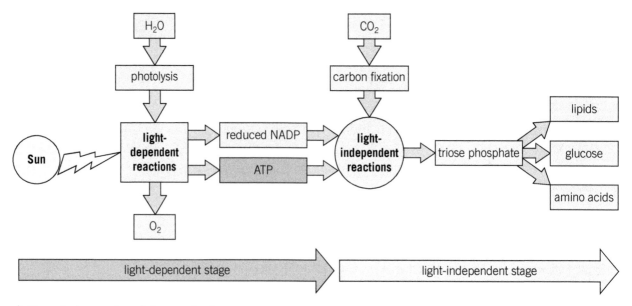

▲ **Figure 3** *An overview of photosynthesis*

Summary questions

1 Name two useful molecules that can be produced from TP and describe one way in which a plant can use each molecule. (*4 marks*)

2 Figure 4 shows a chromatogram of two chloroplast pigments. Calculate R_f values for pigments A and B. (*2 marks*)

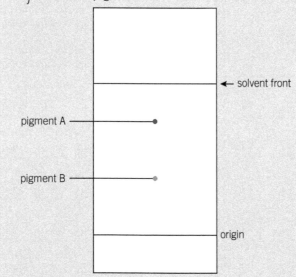

▲ **Figure 4** *Chromatogram of chloroplast pigments*

3 Explain why the Calvin cycle will stop after a plant has been placed in the dark for a period of time. (*2 marks*)

4 One glucose molecule can be produced for every six turns of the Calvin cycle. The mass of one molecule of RuBP is approximately 1.7 times greater than the mass of one molecule of glucose. Estimate the mass of RuBP required to produce 10 kg of glucose. Give your answer to three significant figures. (*2 marks*)

5 Suggest three reasons why plants cannot use the ATP produced in the light-dependent stage of photosynthesis as their only source of ATP. (*3 marks*)

17.4 Factors affecting photosynthesis

Specification reference: 5.2.1(g)i-ii

Several environmental factors influence the rate of photosynthesis. In this topic you will consider some of these limiting factors and the methods that are available to study them.

Limiting factors in photosynthesis

Limiting factor	How does it affect photosynthetic rate?
Light intensity	Light intensity determines the rate at which ATP and reduced NADP are produced in the light-dependent reactions.
Carbon dioxide concentration	Low CO_2 concentrations slow the rate of GP formation.
Temperature	Low temperatures limit the kinetic energy of molecules involved in photosynthetic reactions. High temperatures can cause enzymes such as RuBisCo to denature.

Summary questions

1 Figure 1 shows the effect of carbon dioxide concentration on RuBP and GP concentrations. Left of the central line = 1.0% carbon dioxide. Right of the central line = 0.003% carbon dioxide. Explain the graph. *(2 marks)*

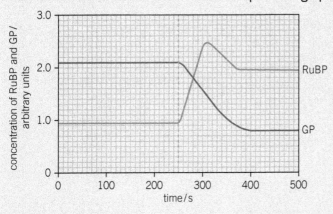

▲ Figure 1

2 Suggest why temperature changes have a greater impact on the rate of the light-independent reactions than the rate of the light-dependent stage of photosynthesis. *(2 marks)*

3 Figure 2 shows the effect of light intensity on GP, TP, and RuBP concentrations. Explain the graph. *(3 marks)*

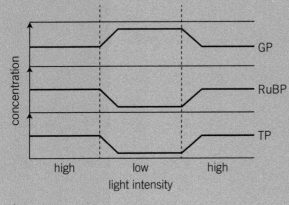

▲ Figure 2

Practical skill: Investigating factors that affect photosynthetic rate

Photosynthesis can be measured using a variety of approaches. **Sensors** and **data loggers** can be used to monitor the effect of various factors on photosynthetic rate.

Several factors can act as independent variables. These include temperature, light intensity, and CO_2 concentration. Remember, in a well-designed experiment only the independent variable will change. All other factors must be controlled. You will therefore be testing the effect of a single factor (the independent variable) on the rate of photosynthesis (the dependent variable).

1 A pigment has an R_f value of 0.60. A student uses paper chromatography. The pigment travels 5.33 cm. How far (to two significant figures) does the solvent in which the pigment is dissolved travel? *(1 mark)*

  A 3.1 cm   **C** 8.8 cm

  B 3.2 cm   **D** 8.9 cm

2 One glucose molecule is produced for every six turns of the Calvin cycle. The mass of one molecule of RuBP is approximately 1.7 times greater than the mass of one molecule of glucose. What is the mass of RuBP, to 2 significant figures, required to produce 5.75 kg of glucose? *(1 mark)*

  A 5800 g   **C** 58 000 g

  B 5900 g   **D** 59 000 g

3 A potometer was used to investigate the effect of the temperature on photosynthetic rate. At 20°C, an oxygen bubble 2.5 cm long was collected over a one-minute period. The radius of the potometer was 0.07 cm. What was the rate of photosynthesis to 2 significant figures? *(1 mark)*

  A 0.038 cm³ min⁻¹   **C** 0.54 cm³ min⁻¹

  B 0.040 cm³ min⁻¹   **D** 0.55 cm³ min⁻¹

3 A potometer was used to investigate the effect of the temperature on photosynthetic rate. At 20°C, an oxygen bubble 2.5 cm long was collected over a one-minute period. The radius of the potometer was 0.07 cm. What was the rate of photosynthesis to 2 significant figures? *(1 mark)*

  A $0.038 \text{ cm}^3 \text{min}^{-1}$ **C** $0.54 \text{ cm}^3 \text{min}^{-1}$

  B $0.040 \text{ cm}^3 \text{min}^{-1}$ **D** $0.55 \text{ cm}^3 \text{min}^{-1}$

4 Discuss the methods that can be employed on farms to increase primary production in crop plants. *(6 marks)*

5 Describe and explain the shape of the graph below. *(3 marks)*

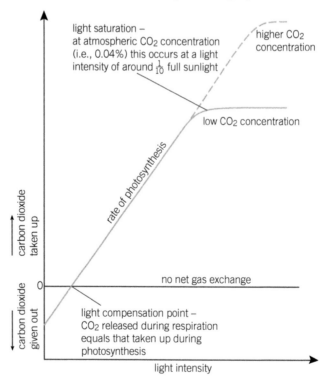

18.1 Glycolysis

Specification reference: 5.2.2(c)

You have learned how energy is trapped in organic molecules by plants. Organic molecules are metabolised to release useable energy in a process called cellular respiration. Here you will examine the first stage of respiration: glycolysis.

> **Revision tip: Step by step ...**
> Glycolysis involves ten steps, but you are required to know only that glucose is converted to hexose bisphosphate, which splits into two TP molecules, which are oxidised to two pyruvate molecules.

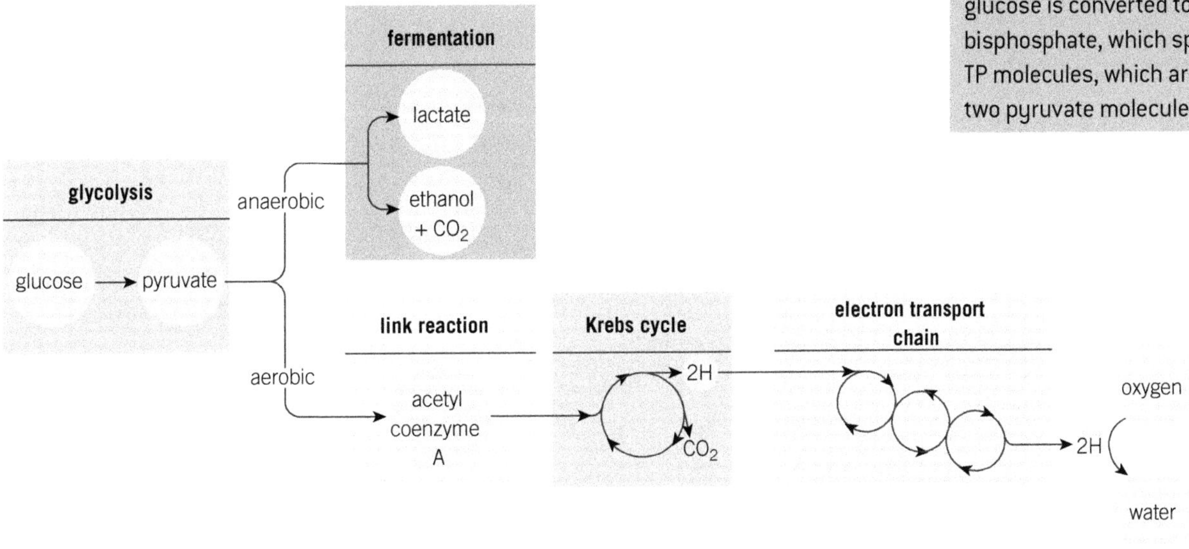

▲ **Figure 1** *An overview of respiration*

Glycolysis

Glycolysis occurs in the **cytoplasm** of cells and comprises the following steps:

1. Glucose is **phosphorylated** to hexose bisphosphate by two **ATP molecules**. This prevents glucose molecules leaving the cell and destabilises them, which makes them easier to break down.
2. Hexose bisphosphate (a six-carbon sugar) **splits** into two molecules of **triose phosphate** (TP – which are three-carbon molecules).
3. TP is **phosphorylated** (by free inorganic phosphate rather than ATP).
4. The two TP molecules lose H atoms (**dehydrogenation**, forming reduced NAD) and phosphate (enabling **ATP formation**) to produce two molecules of **pyruvate**.

▼ **Table 1** *A summary of reactants and products in glycolysis (per glucose molecule)*

Used	Produced
Glucose **2 ATP**	**2 pyruvate molecules** *(which are either used in the link reaction in mitochondria or fermented to enable anaerobic respiration)* **4 ATP** *(through substrate-level phosphorylation)* **2 reduced NAD** *(which are used in the electron transport chain)*

Summary questions

1. Explain why one glucose molecule yields two molecules of ATP from glycolysis. *(3 marks)*
2. Using information in Topic 18.4, Oxidative phosphorylation, estimate the maximum number of ATP molecules that can be generated from one glucose molecule passing through glycolysis. Explain the source of the ATP molecules. *(4 marks)*
3. Describe the role of phosphorylation in glycolysis. *(4 marks)*

> **Synoptic link**
> TP is a molecule produced in the Calvin cycle, which you encountered in Topic 17.3, Photosynthesis. You will learn about the importance of glycolysis for anaerobic respiration in Topic 18.5, Anaerobic respiration.

18.2 Linking glycolysis and the Krebs cycle

Specification reference: 5.2.2(b) and (d)

The three-carbon pyruvate molecules produced in glycolysis need to be converted to two-carbon molecules before they can enter the Krebs cycle. This conversion is known as the link reaction.

The link reaction

The link reaction occurs in the **matrix** of mitochondria. This requires pyruvate to move from the cytoplasm into mitochondria through active transport. Three things happen in the link reaction:

- Pyruvate loses hydrogen (*dehydrogenation*), which produces reduced NAD.

- Pyruvate loses carbon (*decarboxylation*), which produces CO_2 and a two-carbon group called acetyl.

- The acetyl group binds to coenzyme A, which produces acetyl coenzyme A (*acetyl coA*).

The role of coenzyme A is to deliver the acetyl group to the next stage of respiration, the **Krebs cycle**.

Mitochondrion structure

outer mitochondrial membrane
separates the contents of the mitochondrion from the rest of the cell, creating a cellular compartment with ideal conditions for aerobic respiration

inner mitochondrial membrane contains electron transport chains and ATP synthase

cristae are projections of the inner membrane which increase the surface area available for oxidative phosphorylation

matrix
contains enzymes for the Krebs cycle and the link reaction, also contains mitochondrial DNA

intermembrane space
Proteins are pumped into this space by the electron transport chain. The space is small so the concentration builds up quickly

▲ **Figure 1** *The structure of a mitochondrion*

▼ **Table 1** *A summary of reactants and products in the link reaction (per glucose molecule)*

Used	Produced
2 pyruvate molecules	2 CO_2
	2 reduced NAD (*which are used in the electron transport chain*)
	2 acetyl coA

Revision tip: Acetyl vs. acetate

The group that binds to coenzyme A is acetyl (CH_3CO), not acetate (CH_3COO^-).

Synoptic link

You learned about coenzymes in Topic 4.4, Cofactors, coenzymes, and prosthetic groups.

Revision tip: Where there's oxidation, there's reduction

The dehydrogenation of pyruvate is an example of a redox reaction. Pyruvate loses hydrogen and is oxidised, and NAD gains hydrogen and is reduced.

Summary questions

1 Describe what happens to pyruvate after it has been produced in glycolysis. *(4 marks)*

2 Explain what happens to the carbon dioxide produced by the link reaction in plant species. *(2 marks)*

3 Suggest why pyruvate is transported into mitochondria after it is produced in glycolysis. *(2 marks)*

18.3 The Krebs cycle

Specification reference: 5.2.2(e) and (f)

Following the link reaction, acetyl CoA enters the Krebs cycle. This cycle comprises a series of reactions that generate ATP and reduced coenzymes (reduced FAD and NAD). The reduced coenzymes enable even more ATP to be produced by oxidative phosphorylation. Carbon dioxide is released as a by-product of the cycle.

The Krebs (citric acid) cycle

As with the link reaction, the reactions of the Krebs cycle occur in the mitochondrial matrix. The key reactions in the cycle are:

1   Acetyl CoA delivers an **acetyl group**.

2   Acetyl (two-carbon) reacts with **oxaloacetate** (four-carbon) to produce **citrate** (six-carbon).

3   Citrate loses H (*dehydrogenation*, forming reduced NAD) and C (*decarboxylation*, forming CO_2). This results in a five-carbon compound being produced.

4   The five-carbon compound is decarboxylated and dehydrogenated further, regenerating oxaloacetate, which is free to react with another acetyl CoA.

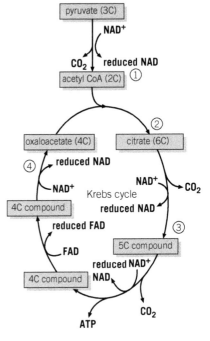

▲ **Figure 1** *The Krebs cycle*

▼ **Table 1** *A summary of reactants and products in the Krebs cycle (per glucose molecule)*

Used	Produced
2 Acetyl-CoA	4 CO_2 **6 reduced NAD** (*which are used in the electron transport chain*) **2 reduced FAD** (*which are used in the electron transport chain*) **2 ATP** (*through substrate-level phosphorylation*)

Revision tip: Just a FAD ... or is it a NAD?

NAD and FAD are two more examples of coenzymes. They both carry protons and electrons from earlier stages of respiration to the electron transport chain. This enables ATP generation. However, as you will see in the next topic, more ATP can be produced from a NAD molecule than a FAD molecule.

Revision tip: Remember these two

Citrate and oxaloacetate are the only Krebs cycle molecules you need to learn. Names of the other intermediates are not required.

Summary questions

1   How many of the original six carbon atoms in a glucose molecule are released as carbon dioxide during the Krebs cycle?   (*1 mark*)

2   Using information in Topic 18.4, Oxidative phosphorylation, estimate the maximum number of ATP molecules that can be generated from one turn of the Krebs cycle. Explain the source of the ATP molecules.   (*4 marks*)

3   Suggest why oxaloacetate is present in cells at very low concentrations.   (*2 marks*)

18.4 Oxidative phosphorylation

Specification reference: 5.2.2(g)

Most of the ATP generated in respiration is produced via oxidative phosphorylation. The reduced coenzymes (FADH$_2$ and NADH) produced during earlier stages of respiration donate H$^+$ ions and electrons to an electron transport chain. ATP is synthesised by chemiosmosis.

The electron transport chain

Electron transport chains (ETCs) are located on inner mitochondrial membranes. The chains use energy from electrons to pump H$^+$ ions into the intermembrane space. A proton gradient is established, which enables chemiosmosis to occur through ATP synthase. The sequence of events is as follows:

1 High energy **electrons** are passed from NADH and FADH$_2$ to **electron carriers** in the ETC.

2 Electrons are passed between carriers in a series of **redox reactions**.

3 Energy is released from each reaction.

4 The energy is used to **pump H$^+$ ions** from the matrix into the intermembrane space.

5 H$^+$ ions diffuse through ATP synthase (**chemiosmosis**) back into the matrix, which produces ATP.

6 H$^+$ ions and electrons react with O$_2$ to produce water.

Revision tip: Just a number...

You are not required to learn how many ATP molecules are generated in oxidative phosphorylation. However, you might be provided with relevant data and asked to interpret the information.

▼ **Table 1** *A summary of reactants and products in oxidative phosphorylation (per glucose molecule)*

Used	Produced
10 reduced NAD (NADH) **2 reduced FAD (FADH$_2$)** **6 O$_2$**	**26–34 ATP** (estimates vary – see Go further for more details) **6 H$_2$O**

Synoptic link

You examined the principles of chemiosmosis in Topic 17.2, ATP synthesis.

Summary questions

1 Explain how the structure of mitochondrial cristae maintains a high rate of oxidative phosphorylation. *(2 marks)*

2 Explain why oxygen is called the final electron acceptor in respiration. *(3 marks)*

3 The theoretical maximum yield of ATP from aerobic respiration may not always be achieved. Suggest why. *(2 marks)*

Go further: How many ATP molecules are produced per glucose molecule?

Scientists have calculated different estimates of the number of ATP molecules produced per glucose molecule. Some suggest 38 is the theoretical maximum yield. However, the current prevailing opinion is that 30 or 32 ATP molecules are produced per glucose. The crux of the calculation is working out how many ATPs are generated for each NADH and FADH$_2$. Most scientists think that NADH produces 2.5 ATPs, whereas FADH$_2$, because it enters the ETC at a later stage, produces only 1.5. The following table summarises current understanding:

Stage of respiration	Source of ATP	Number of ATP molecules
Glycolysis	Substrate-level phosphorylation	2
	2 NADH (oxidative phosphorylation)	3–5
Link reaction	2 NADH (oxidative phosphorylation)	5
Krebs cycle	Substrate-level phosphorylation	2
	6 NADH (oxidative phosphorylation)	15
	2 FADH$_2$ (oxidative phosphorylation)	3
Total ATP yield (per glucose)		30–32

1 Suggest why only 1.5 ATP molecules might be produced from a molecule of reduced NAD generated during glycolysis.

2 Use the following information to determine where reduced NAD and reduced FAD enter the electron transport chain:

Four protein carriers are present. Carrier 1 and 3 each pump 4 H$^+$ ions per coenzyme. Carrier 2 pumps none, and carrier 4 pumps 2 H$^+$ ions into the intermembrane space. Scientists think that 4 H$^+$ ions are required to pass through ATP synthase to synthesise one molecule of ATP.

18.5 Anaerobic respiration

Specification reference: 5.2.2(i)

Oxygen, as you learned in the previous topic, is the electron acceptor at the end of respiration. ATP cannot be produced via oxidative phosphorylation without oxygen. However, organisms can still generate some ATP even when oxygen is unavailable.

Anaerobic respiration

Anaerobic respiration generates small amounts of ATP through **substrate-level phosphorylation** in **glycolysis**. The link reaction, the Krebs cycle, and the electron transport chain all stop working in the absence of oxygen.

Fermentation

The function of fermentation is to oxidise NADH, thereby regenerating NAD. This enables glycolysis to continue. Glycolysis would stop without fermentation because supplies of NAD would run out, and TP would no longer be converted to pyruvate.

Fermentation in **mammals** involves reduced NAD donating H atoms to pyruvate, producing lactate:

Pyruvate + reduced NAD → lactate + NAD

Fermentation in **yeast** and **plants** involves pyruvate breaking down to ethanal and CO_2. Reduced NAD then donates H atoms to ethanal, producing ethanol:

Pyruvate → ethanal + CO_2   *Ethanal + reduced NAD → ethanol + NAD*

> **Key term**
>
> **Fermentation:** The breakdown of pyruvate into lactate or ethanol.

> ### Practical skill: Measuring anaerobic respiration rates
>
> The rate of anaerobic respiration in yeast can be investigated using the following method:
>
> - Yeast cultures are placed in sealed flasks (*to ensure anaerobic conditions*).
>
> - The **dependent variable** is the rate of carbon dioxide production (*which can be measured by monitoring the displacement of coloured liquid in a capillary tube; this is similar to the use of a potometer, which you learned about in Topic 9.3, Transpiration*).
>
> - When testing the effect of an **independent variable** (e.g. temperature) on respiration rate, only that variable should be changed. All other factors (e.g. pH, glucose concentration) are **controlled.**

> **Revision tip: Respiration efficiency**
>
> Anaerobic respiration is far less efficient than aerobic respiration. One molecule of glucose can produce 32 ATP molecules in aerobic respiration but only two ATP molecules in anaerobic respiration.

Summary questions

1 Explain why mammals use anaerobic respiration only for short periods of time. *(2 marks)*

2 The hydrolysis of one mole of glucose releases 2880 kJ of energy. The hydrolysis of one mole of ATP releases 30.6 kJ of energy. For every mole of glucose, aerobic respiration produces 32 moles of ATP, whereas anaerobic respiration produces 2 moles of ATP. Calculate the percentage efficiencies of the two forms of respiration. *(2 marks)*

3 Suggest why the absence of oxygen stops
   a chemiosmosis and the electron transport chain *(2 marks)*
   b the Krebs cycle. *(2 marks)*

18.6 Respiratory substrates

Molecules other than glucose can be used as respiratory substrates. Amino acids, glycerol, and fatty acids can all be metabolised and enter respiration at different points.

Respiratory substrates

The following table outlines how alternative respiratory substrates are used.

Respiratory substrate		Which stage of respiration does it enter?	Which molecule is it used to form?
Amino acids		The link reaction or the Krebs cycle	This depends on the amino acid e.g. Glycine → pyruvate Isoleucine → acetyl CoA Aspartate → oxaloacetate
Triglycerides	Glycerol	Glycolysis	Triose phosphate
	Fatty acids	Krebs cycle	Acetyl CoA
Lactate		Link reaction	Pyruvate

Worked example: Respiratory quotient

Respiratory quotient $(RQ) = \dfrac{\text{Carbon dioxide (produced)}}{\text{Oxygen (consumed)}}$

How do you work out RQ for a particular respiratory substrate?

Let us examine the respiration of stearic acid $(C_{17}H_{35}COOH)$ as an example.

1 **Write an equation.** $C_{17}H_{35}COOH + (?)O_2 \rightarrow (?)CO_2 + (?)H_2O$

2 The number of CO_2 = number of C atoms in the substrate.
$C_{17}H_{35}COOH + (?)O_2 \rightarrow \mathbf{18}CO_2 + (?)H_2O$

3 The number of H_2O = half the H atoms in the substrate.
$C_{17}H_{35}COOH + (?)O_2 \rightarrow 18CO_2 + \mathbf{18}H_2O$

4 Balance the number of oxygen atoms on each side.
$C_{17}H_{35}COOH + \mathbf{26}O_2 \rightarrow 18CO_2 + 18H_2O$

5 Calculate RQ: $18/26 = 0.69$.

Revision tip: What does RQ tell us?

RQ of more than 1 = some carbon dioxide is produced from anaerobic respiration

RQ of 1 or less = aerobic respiration

An RQ of 1 indicates carbohydrates are being respired, 0.8–0.9 indicates proteins, and approximately 0.7 indicates lipids.

Summary questions

1 What must be removed from all amino acids before they can be metabolised as respiratory substrates? State the name of this process. *(2 marks)*

2 Calculate the RQ of palmitic acid $(C_{15}H_{31}COOH)$. Give your answer to three significant figures. *(3 marks)*

3 Some respiratory molecules are easier than others to form from an amino acid. Suggest the likely respiratory molecules formed from the following amino acids. Explain your answers.
   a asparagine $(C_4H_8N_2O_3)$
   b alanine $(C_3H_7NO_2)$. *(4 marks)*

Chapter 18 Practice questions

1. Which of the following statements is/are true of the conversion of pyruvate to acetyl coenzyme A in the link reaction? *(1 mark)*

   1. Pyruvate is reduced.
   2. Pyruvate is decarboxylated.
   3. Pyruvate is dehydrogenated.

   A 1, 2, and 3 are correct

   B Only 1 and 2 are correct

   C Only 2 and 3 are correct

   D Only 1 is correct

2. Which of the following statements is/are true of the electron transport chain in mitochondria? *(1 mark)*

   1. The first electron carrier is reduced by NADH.
   2. Oxygen acts as the terminal electron acceptor.
   3. Energy is used to pump protons out of the intermembrane space.

   A 1, 2, and 3 are correct

   B Only 1 and 2 are correct

   C Only 2 and 3 are correct

   D Only 1 is correct

3. Which molecule acts as the hydrogen acceptor during alcoholic fermentation? *(1 mark)*

   A Glucose   **C** Ethanol

   B Pyruvate   **D** Ethanal

4. What is the respiratory quotient of palmitic acid ($CH_3(CH_2)_{14}COOH$) to two significant figures? *(1 mark)*

   A 0.69

   B 0.70

   C 0.71

   D 0.72

5. Describe how the respiratory quotient of a plant can be determined using a respirometer. *(6 marks)*

6. The graph below illustrates the release of energy during respiration. Explain the difference between respiration and combustion in terms of the release of chemical energy and its significance. *(3 marks)*

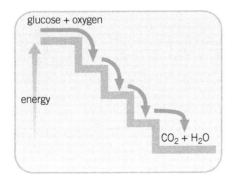

Changes to the base sequences in DNA are known as **mutations**. These alterations vary in their severity. They can be harmful, beneficial, or have no effect.

Types of mutation

You can consider mutations in terms of how the DNA changes or the effect of the mutation.

Change to DNA	Description	Examples of genetic diseases
Substitution	One nucleotide is exchanged for another.	Sickle cell anaemia
Insertion	An extra nucleotide (or more than one) is placed in the DNA sequence.	Huntington's disease
Deletion	A nucleotide (or more than one) is removed from the DNA sequence.	Cystic fibrosis

Revision tip: Frameshifts

Substitution mutations change one codon and cause a maximum of one amino acid to be altered in a protein's primary structure. Insertion and deletion mutations can change many codons along the DNA sequence; this is called a **frameshift** mutation. These mutations tend to be much more harmful than substitution mutations.

Revision tip: Remaining neutral?

Some mutations change the primary structure of a protein but have no significant effect on its function (i.e. they have a neutral effect).

Type of mutation	Description	Effects
Silent	The mutation could: • be in a **non-coding region** • produce a new codon that codes for the same amino acid (because of the **degenerate code**).	No effect (neutral)
Nonsense	A codon is changed to a **stop codon**.	A shorter polypeptide Normally harmful
Missense	At least one new amino acid is introduced into a protein's primary structure.	Harmful usually, but sometimes beneficial or neutral

Revision tip: Evolution requires mutation

You learned in Topic 10.8, Changing population characteristics, that natural selection requires variation within a population. This variation can be attributed to DNA mutations, which produce variants of genes (i.e. alleles). Occasionally a slight change in a gene can be beneficial. It may improve the function of the polypeptide for which it codes. These slight alterations to DNA sequences tend to be substitution mutations rather than insertions or deletions.

Go further: The FTO 'hunger' gene: an example of the subtle effects of gene mutation?

The FTO gene codes for a protein that regulates appetite and feeding behaviour. Mutations have produced several known variants of the gene. One of these alleles has been linked to an increased risk of obesity.

In a study, people who are homozygous for the risk allele were found to be 3 kg heavier on average and were 70% more likely to have obesity than people without a copy of this allele. A later study found that people with the risk allele have higher levels of the 'hunger hormone' ghrelin in their blood, which means they begin to feel hungry sooner than other people. Ghrelin is coded for by a separate gene.

1 Suggest the type of mutation responsible for the FTO gene variants. Explain your reasoning.

2 Suggest how the protein produced by the FTO gene might control feeding behaviour.

Synoptic link
You looked at the structure of DNA and the degeneracy of the genetic code in Chapter 3, Biological molecules.

Revision tip: Mutagens
DNA mutations can occur spontaneously, but the rate of mutation is increased by mutagens, which can be biological (e.g. viruses), chemical , or physical (e.g. radiation).

Summary questions

1 Explain why an insertion mutation of three nucleotides does not cause a frameshift.   (*3 marks*)

2 Look at the base sequence below. How many triplet codes in this sequence would be changed by
   a all guanine bases experiencing substitution mutations?
   b a substitution mutation of A?
   c a deletion of A?   (*3 marks*)
   G C A C T T G G G C C T C

3 Aromatic rice is a popular cooking ingredient. A 'fragrance' gene has been sequenced in rice plants and found to have two variants: one that results in an aroma and one that does not. Sections of the two gene variants are shown below. The fragrant variant has four separate mutations that make it different to the non-fragrant variant. Identify and describe the four mutations.   (*4 marks*)
   Non-fragrant variant: A A A C T G G T A A A A A G A T T A T G G C T T C A G C T G
   Fragrant variant: A A A C T G G T A T A T A T T T C A G C T G

19.2 Control of gene expression

Specification reference: 6.1.1(b)

A gene is expressed when its sequence of codons is transcribed and translated into a polypeptide. Gene expression can be controlled in several ways, including the prevention of transcription and the post-transcriptional modification of mRNA. Even after translation, a polypeptide can be modified to change its function.

How is gene expression controlled?

The table below outlines how gene expression can be controlled at different stages of polypeptide production.

Stage	Control mechanism	Description
Transcription	Chromatin structural changes	**Heterochromatin** forms during cell division. DNA is wound tightly around histones. No transcription occurs.
		Euchromatin forms during interphase. DNA is wound loosely around histones. Transcription can occur.
	Transcription factors	Molecules that bind to DNA to either promote or prevent transcription.
	Epigenetics	**Acetylation** of histones increases transcription rates.
		Methylation of DNA prevents transcription.
	Operons (e.g. the *lac operon* in prokaryotes, see Figure 1)	Genes are switched *off* when a **repressor** binds to an **operator** region, which blocks a **promoter** region. This prevents **RNA polymerase** binding and stops the transcription of **structural genes**.
		Genes are switched *on* when the repressor is removed.
Post-transcription	mRNA processing	**Splicing**: introns (non-coding DNA) are removed from mRNA. Different polypeptides can be formed by retaining some **introns** and rearranging **exons**.
	mRNA editing	mRNA can be edited by adding, deleting, or substituting nucleotides.
Translation	Control of mRNA binding	**Inhibitory proteins** prevent the binding of mRNA to ribosomes. **Initiation factors** promote mRNA binding.
Post-translation	Polypeptide modification	For example, protein folding, and the addition of non-protein groups and disulfide bridges.

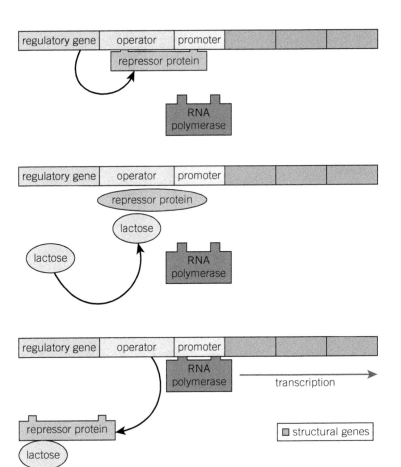

▲ **Figure 1** *The lac operon*

Synoptic link

You learned about transcription and translation in Topic 3.10, Protein synthesis.

Revision tip: Structural vs. regulatory

Regulatory genes code for proteins that control the expression of other genes (e.g. repressor proteins). Structural genes code for proteins that are not involved in regulation (i.e. most of the proteins that you have learned about, such as enzymes and hormones).

Summary questions

1 Explain how a gene consisting of 930 base pairs can be transcribed into mature mRNA consisting of only 615 nucleotides. (*2 marks*)

2 Explain the importance of splicing in organisms. (*2 marks*)

3 Figure 2 shows some different conditions shared by twins, and illustrates the relative influence of genes and the environment on each trait. Explain the evidence for the comparative influence of genetics and the environment on height and strokes. (*3 marks*)

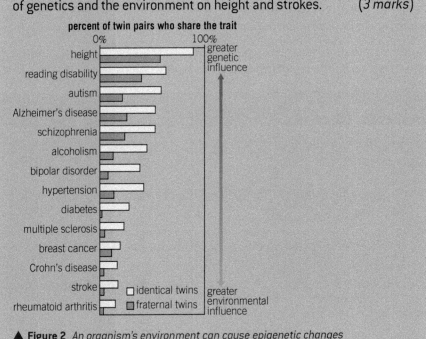

▲ **Figure 2** *An organism's environment can cause epigenetic changes to genes. This helps to explain phenotypic differences between twins*

19.3 Body plans

An organism's development is governed by sets of regulatory genes. Fungi, plants, and animals all possess these DNA sequences, which are known as **homeobox genes**. The development of an organism's body plan is also reliant on the balance between **apoptosis,** otherwise known as programmed cell death, and mitosis.

Homeobox genes

Question and model answer: Homeobox genes

Q. Describe the role of homeobox genes in the development of eukaryotic organisms.

A.

- Homeobox genes are regulatory genes. Homeobox sequences within these genes are 180 base pairs long and code for polypeptide sequences that are 60 amino acids long. These polypeptides are called **homeodomains**.

- Homeodomains bind to DNA and switch genes on or off. They are transcription factors.

- Hox genes are a subset of homeobox genes that are present only in animals. Species possess different numbers of these genes (e.g. vertebrates have four clusters of Hox genes).

- Homeobox genes are expressed in a set order.

- By regulating which genes are expressed in different parts of an organism, homeobox genes control the development of the organism's body and ensure structures develop in the correct positions.

- For example, a role of homeobox genes early in development is to determine the tail and head regions of an organism (i.e. its polarity).

- Homeobox genes regulate both mitosis and apoptosis.

Apoptosis

Apoptosis is the mechanism the body employs to destroy cells in a controlled fashion. It always involves the same steps: the cell shrinks, the nucleus condenses, enzymes break down the cytoskeleton, the cell breaks into fragments held in vesicles, and macrophages digest the cell fragments.

In addition to destroying damaged cells, apoptosis is vital in development. It prunes excess cells and sculpts structures in developing organisms.

Synoptic link

You can remind yourself of how mitosis operates by reading Topic 6.2, Mitosis.

Revision tip: Mitosis vs. apoptosis

Remember that in organs and tissues both apoptosis and mitosis will be happening. The two processes are balanced if a region of tissue or an organ remains the same size. The organ will decrease in size if the rate of apoptosis exceeds mitosis.

Summary questions

1   Describe two examples of apoptosis in an organism's development.
(2 marks)

2   Suggest why species differ in the number of homeobox genes they possess.
(2 marks)

3   Necrosis is a damaging form of cell death caused by infection or trauma. Necrotic cells rupture and release hydrolytic enzymes. Outline and explain the differences between apoptosis and necrosis. *(4 marks)*

1   What is the name given to the type of mutation that results in the
    formation of a stop codon?   (*1 mark*)

    A Nonsense   **C** Silent

    B Missense   **D** Insertion

2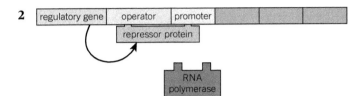

    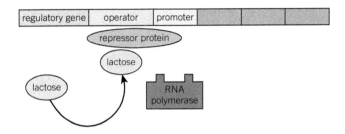

    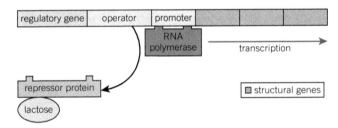

    a What is the overall name given to the DNA regions shown in the
       diagram?   (*1 mark*)

    b Describe the role of lactose in the process shown in the diagram.
                                                         (*2 marks*)

3   Complete the following passage about apoptosis by adding the most
    appropriate words or phrases to the gaps.   (*4 marks*)

    Apoptosis is also known as .... cell death. A cell's
    cytoskeleton is broken down by .... The cell shrinks
    and the plasma membrane forms bulges called ....
    The cell fragments and apoptotic bodies form, which are engulfed by
    ....

4   Explain how the process illustrated in the diagram below contributes to
    the control of gene expression.   (*3 marks*)

5   Match each description of a mutation with the correct type of
    mutation. Choose from: missense, point, silent.   (*3 marks*)

    A Thymine is exchanged for cytosine in the base sequence of DNA.

    B A mutation occurs in a non-coding region of DNA.

    C Three new amino acids are introduced into the primary structure of
       a protein.

20.1 Variation and inheritance

Specification reference: 6.1.2(a) and (d)

Synoptic link

You learned about the environmental and genetic causes of variation in Topic 10.5, Types of variation. Topic 10.6, Representing variation graphically, outlined the differences between continuous and discontinuous variation.

Phenotypic variation within species can have both environmental and genetic origins. The relative influence of the environment and genes determines whether a characteristic exhibits discontinuous or continuous variation.

What causes phenotypic variation?

Genetics

Members of a species can possess different versions of genes (i.e. alleles). These gene variants are produced by DNA **mutations**. Sexual reproduction shuffles alleles, producing new allele combinations in gametes through the following processes:

- crossing over (during meiosis)
- independent assortment (during meiosis)
- random fusion of gametes during fertilisation.

The different alleles in a population code for different polypeptides, which establishes phenotypic variation.

Environment

Characteristics such as eye colour and gender are unlikely to change during an individual's lifetime. However, many other traits can be sculpted by the environment. For example, body mass, while in part determined by genetics, is influenced by nutritional intake and activity levels. The length of a plant's leaves is determined by mineral intake, ambient temperatures, and its exposure to light.

Continuous vs. discontinuous variation

When both genetics and the environment play roles in shaping a characteristic, individuals exhibit continuous variation. Discontinuous variation is a result of genetics alone determining a characteristic.

Type of variation	Definition	Cause of variation	How is the data displayed?	Examples
Discontinuous	A characteristic that has specific (discrete) values (without intermediate values)	Genetics (one or two genes)	Bar graph (usually qualitative data)	Blood groups Genetic diseases (i.e. someone either has cystic fibrosis or does not)
Continuous	A characteristic that has any value within a range	Environment and genetics (usually several genes)	Line graph (quantitative data)	Body mass Height Blood glucose concentration

Revision tip: Continuous or discontinuous?

Deciding whether a characteristic shows continuous or discontinuous variation is not always straightforward. For example, we can divide eye colours into discrete categories: brown, green, blue. However, in reality, eyes come in many shades of blue, green, and brown. These colours can be measured precisely and presented in a line graph to show continuous data.

Summary questions

1 Which of the following statements represents continuous variation and which represents discontinuous variation?
   a A characteristic controlled by one gene with five different alleles.
   b Quantitative data presented in a line graph.
   c A person's lung capacity. *(3 marks)*

2 A tomato plant was cloned. One cloned plant was grown at 10°C and another was grown at 20°C. Suggest and explain the effect on the size of the tomato fruit. *(3 marks)*

3 Identical twins are sometimes separated early in life and are raised in different environments. Suggest how these sets of twins can be used to assess the causes of variation in human populations. *(3 marks)*

20.2 Monogenic inheritance
Specification reference: 6.1.2(b)(i)

Patterns of inheritance can be illustrated with genetic diagrams, which show the probability of offspring inheriting certain alleles from their parents. Here you will learn about monogenic inheritance (i.e. inheritance patterns involving one gene), including the inheritance of codominant alleles, multiple alleles, and sex-linked genes.

Genetic crosses
Genes with one recessive allele and one dominant allele
Genetic diagrams can be drawn to illustrate the probability of inheriting a particular genotype.

 Worked example: Cystic fibrosis

Cystic fibrosis (CF) is a genetic disease with two alleles: a dominant healthy allele (F) and a faulty recessive allele (f).

Example 1: A genetic cross between a homozygous dominant father and a homozygous recessive mother.

Step 1: State the genotypes of both parents (FF and ff in this example).

Step 2: State the possible gametes that each parent could pass on, and draw a Punnett square to show the possible genotypes of their offspring.

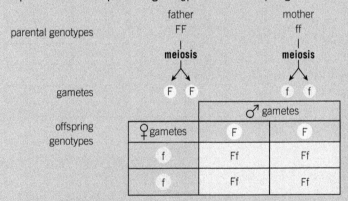

Step 3: State the probability of offspring having particular phenotypes.

In this example 100% of the offspring will be healthy carriers of the cystic fibrosis allele.

Example 2: Two heterozygous parents.

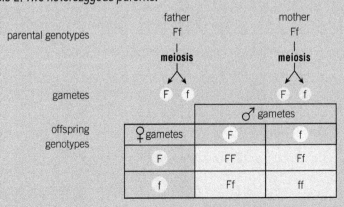

In this example there is a 25% probability that an offspring will have cystic fibrosis (genotype ff).

Revision tip: Important cases...
By convention, dominant alleles are represented by upper case letters. It is good practice to choose a letter where the upper and lower case are clearly different. So Ff is a better choice than Cc to represent the dominant and recessive cystic fibrosis alleles.

Revision tip: Alternative expressions
The results of genetic crosses can be expressed in different forms. e.g. 2 in 4 probability = 50% = 2:2 ratio of phenotypes.

Revision tip: Representing codominance
Codominant alleles cannot be represented by a combination of upper and lower case letters. Instead, one letter is chosen to represent the gene, and two other letters are written in superscript to represent the codominant alleles. For example, the normal allele for haemoglobin production is written as H^A, and the codominant allele for sickle cell haemoglobin is H^S.

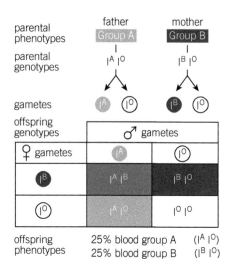

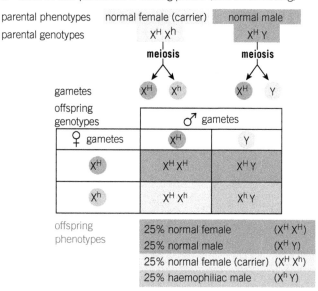

offspring phenotypes
- 25% blood group A (I^A I^O)
- 25% blood group B (I^B I^O)
- 25% blood group AB (I^A I^B)
- 25% blood group O (I^O I^O)

▲ **Figure 1** *The inheritance of the ABO blood group gene, an example of a gene with multiple alleles*

Revision tip: Representing sex linkage

When displaying sex-linked genotypes, an appropriate letter is chosen to represent the gene (e.g. 'H' for haemophilia). Lower and upper cases are used for recessive and dominant alleles. This allele letter is placed as superscript on the X or Y chromosome (e.g. X^H and X^h).

Codominance

Both alleles are expressed when a genotype consists of two codominant alleles.

For example, sickle cell anaemia is a disease in which a faulty version of the haemoglobin molecule is produced. The genotype H^A H^A results in normal haemoglobin production, whereas H^S H^S results in the disease. The heterozygous genotype H^A H^S produces both versions of haemoglobin because the alleles show codominance.

Multiple alleles

Many genes have more than two possible versions. These multiple gene variants are a mixture of dominant, codominant, and recessive alleles. The gene for the ABO blood group, for example, has two codominant alleles (I^A and I^B) and one allele that is recessive to them both (I^O).

Sex linkage

Genes located on the X or Y chromosomes are called sex-linked. Two aspects of sex linkage are:

- more genes are found on the X chromosome because it is larger than the Y chromosome
- males are more likely to suffer from a sex-linked disease because they inherit one X chromosome. For recessive conditions, only one recessive allele would be required for them to inherit the disease.

offspring phenotypes
- 25% normal female (X^H X^H)
- 25% normal male (X^H Y)
- 25% normal female (carrier) (X^H X^h)
- 25% haemophiliac male (X^h Y)

▲ **Figure 2** *Haemophilia inheritance, an example of sex linkage*

Key terms

Genotype: The genetic composition of an organism, which describes all the alleles it contains. Genotypes for a particular locus can be heterozygous or homozygous.

Phenotype: An organism's observable characteristics.

Homozygous: Having two identical alleles of a gene. Homozygous genotypes can be recessive or dominant.

Heterozygous: Having two different alleles of a gene.

Summary questions

1 Suggest why more X-linked traits exist than Y-linked traits. *(2 marks)*

2 The colour of the four o'clock flower (*Mirabilis jalapa*) is controlled by two codominant alleles: C^R (red flowers) and C^W (white flowers). What are the genotypes of the two plants that would need to be bred to produce 50% red offspring and 50% pink offspring? *(2 marks)*

3 Paul and Sandra have a child together. Neither has cystic fibrosis, but Sandra is a carrier (genotype Ff). Paul has the genotype FF. Their child, many years later, has a baby whose other parent is a carrier of cystic fibrosis. Calculate the probability that Paul and Sandra will have: **a** a child with cystic fibrosis **b** a child who is a carrier of cystic fibrosis **c** a grandchild with cystic fibrosis **d** a grandchild who is a carrier of cystic fibrosis. *(4 marks)*

20.3 Dihybrid inheritance

Specification reference: 6.1.2(b)(i)

A dihybrid cross shows the inheritance pattern of two characteristics, coded for by two different genes.

Genetic cross with two genes

Genetic diagrams can be drawn to illustrate dihybrid inheritance. These diagrams have a similar format to the monohybrid crosses in Topic 20.2, Monogenic inheritance, but at least four possible alleles exist (two per gene) and therefore four potential phenotypes exist.

 Worked example: *Drosophila* eyes and wings

Two observable traits in *Drosophila* flies are eye colour and wing size. The allele for red eyes is R (dominant). The recessive allele for white eyes is r (recessive). Wing shapes can be normal (W, a dominant allele) or vestigial (w, a recessive allele). The possible genotypes of offspring from two heterozygous parents (RrWw) are shown in the genetic diagram below.

Male gametes	Female gametes			
	RW	Rw	rW	rw
RW	RRWW	RRWw	RrWW	RrWw
Rw	RRWw	RRww	RrWw	Rrww
rW	RrWW	RrWw	rrWW	rrWw
rw	RrWw	Rrww	rrWw	rrww

The phenotypes of offspring will be present in the following ratios:

9 red eyes, normal wings

3 white eyes, normal wings

3 red eyes, vestigial wings

1 white eyes, vestigial wings

A scientist can assess whether their predictions about the genetics of a species are correct by using the chi-squared test, which you will learn about in Topic 20.4, Phenotypic ratios. If the phenotypic ratio for a dihybrid cross differs from what is expected then other effects, such as epistasis and linkage, might be present.

Revision tip: Phenotypic ratios

Keep an eye out for certain phenotypic ratios in offspring. For example, if two heterozygous parents are used for a dihybrid cross then the expected offspring phenotypes will be in a 9:3:3:1 ratio.

Summary questions

1. A white-eyed male *Drosophila* with vestigial wings had offspring that were all red-eyed with normal wings. Predict the genotype of the mother. *(2 marks)*

2. A scientist studied the inheritance of two traits in pea plants: seed colour and seed texture. The alleles for these traits are: R = round seeds, r = wrinkled seeds, G = yellow seeds, g = green seeds. R and G are the dominant alleles. What proportion of each phenotype would you expect to find in the offspring of a plant with the genotype RRGg and another plant that is RrGg? *(2 marks)*

3. When predicting a phenotypic ratio of 9:3:3:1 for a dihybrid cross, what assumptions do you have to make about the two gene loci being studied? *(2 marks)*

20.4 Phenotypic ratios

Specification reference: 6.1.2(b)(ii) and (c)

Simple inheritance patterns can be predicted when two genes are inherited independently. Sometimes, however, two genes are connected in some way. The **chi-squared (χ^2) test** compares actual offspring phenotypes against the expected phenotypic ratios. If the offspring phenotypic ratios differ from what is expected then this suggests the genes are not inherited independently. For example, two genes can be located on the same chromosome (**autosomal linkage**), or one gene can affect the expression of another (**epistasis**).

Autosomal linkage

Two genes are said to be linked when they are located on the same autosome (i.e. a chromosome other than X or Y). This is important because:

- Linked allele combinations will be inherited together (as a single unit).
- Only crossing over during meiosis (see Topic 6.3, Meiosis) can separate linked allele combinations.
- The nearer two genes are on a chromosome, the less likely they are to be separated during crossing over.

> ### Synoptic link
>
> You learned about sex linkage in Topic 20.3, Dihybrid inheritance. Do not confuse this with autosomal linkage.

Epistasis

Epistasis occurs when one gene prevents the expression of another gene. Two forms of epistasis exist: recessive and dominant.

	Description	Example
Recessive epistasis	The epistatic gene (i.e. the gene doing the suppressing) needs to be **homozygous recessive** to prevent the expression of the other gene.	Flower colour in *Salvia* is controlled by two genes. Dominant allele B = purple flowers Recessive allele b = pink flowers However, when another gene is homozygous recessive (aa), the flowers are white. The genotype of B/b becomes irrelevant. AaBB, AABB, AABb, AaBb = purple flowers Aabb, AAbb = pink flowers aaBB, aaBb, aabb = white flowers
Dominant epistasis	The epistatic gene needs **at least one dominant allele** to prevent the expression of the other gene.	Fruit colour in summer squash is also controlled by two genes. Dominant allele E = yellow fruit Recessive allele e = green fruit However, the presence of one dominant D allele at another gene results in white fruit. ddEe, ddEE = yellow fruit ddee = green fruit DdEE, DdEe, Ddee, DDEE, DDEe, DDee = white fruit

Chi-squared test

The χ^2 test assesses whether there is a significant difference between the observed and expected numbers of offspring phenotypes.

$$\chi^2 = \Sigma \frac{(O-E)^2}{E} = \text{the sum of } \frac{(\text{observed numbers} - \text{expected numbers})^2}{\text{expected numbers}}$$

Model question and answer: Calculating chi-squared

The results from the *Drosophila* genetic cross shown in Topic 20.3, Dihybrid inheritance, can be analysed using the χ^2 test.

Q. Calculate χ^2 for this genetic cross and determine whether the predicted model of inheritance can be accepted.

The expected values are based on the predicted phenotypic ratios of the offspring (see Topic 20.3, Dihybrid inheritance).

A.

Phenotype	Observed (O)	Expected (E)	O – E	(O – E)²	(O – E)²/E
Red eyes, normal wings	87	90	–3	9	0.100
White eyes, normal wings	31	30	1	1	0.033
Red eyes, vestigial wings	35	30	5	25	0.833
White eyes, vestigial wings	7	10	–3	9	0.900
					$\chi^2 = 1.87$

The next stage is to compare the calculated value to a χ^2 significance table.

The degrees of freedom are calculated as the number of categories (phenotypes) – 1. In this example we have 3 degrees of freedom.

The *P* value tells us the probability of differences being a result of chance. We say that differences are statistically significant if $P = 0.05$ or less (i.e. $P = 0.05$ would mean there is only a 5% probability that differences can be attributed to chance). In this case, the critical value of χ^2 (which represents a *P* value of 0.05) would be 7.81.

In our example, the P value for a calculated χ^2 of 1.87 lies somewhere between 0.5 and 0.9. We can be very confident that the differences between our observed and expected results are down to chance and are not significant. This means we can accept our model of inheritance for these two traits.

Key terms

Autosomal linkage: Genes located on the same (non-sex) chromosome.

Epistasis: The effect of one gene on the expression of another.

Revision tip: More phenotypic ratios

Remember that a 9:3:3:1 offspring phenotypic ratio is expected if both parents are heterozygous for two unconnected genes.

Linkage will alter this ratio. Many different ratios can occur, depending on the alleles that are linked and how closely they are linked.

Dominant epistasis is likely to produce a 12:3:1 or 13:3 ratio (with heterozygous parents).

Recessive epistasis will produce a 9:3:4 offspring phenotypic ratio (with heterozygous parents).

Summary questions

1 Humans with red hair tend to have pale skin and green eyes. Suggest a reason for this inheritance pattern. (*2 marks*)

2 Pea plants have the following alleles for seed colour and texture: R = round, r = wrinkled, G = yellow, g = green.
Two pea plants were bred together to produce 100 offspring. A scientist expected 50 plants that produced round, yellow seeds and 50 that produced wrinkled, yellow seeds. She expected no green seeds. The results were: 55 round, yellow seeds; 45 wrinkled, yellow seeds; 0 round, green seeds; 0 wrinkled, green seeds. Calculate whether the scientist's predictions can be supported by the test. The critical value of χ^2 in this case is 3.84. Suggest the likely genotypes of the two plants being bred. (*4 marks*)

3 Suggest a likely molecular mechanism for
   a recessive epistasis
   b dominant epistasis. (*4 marks*)

20.5 Evolution

Evolution occurs when genetic mutations and natural selection result in a change of allele frequencies in a population. In theory, allele frequencies remain constant in a stable population. These allele frequencies can be calculated using Hardy–Weinberg equations, which are discussed here. When a population decreases in size, however, allele frequencies are liable to change.

Factors affecting allele frequencies
Forms of natural selection

A selection pressure can alter the distribution of phenotypes in a population. The selection of phenotypes can be stabilising or directional.

Form of selection	Description	Appearance (arrows show selection pressure)	Example
Stabilising	Selection favours **average phenotypes**. Alleles that produce extreme traits are eliminated.	evolved population original population	Most mammals will have fur length close to the mean in an environment with a stable temperature. Individuals with short or long fur are less likely to survive and reproduce. These alleles are therefore eliminated.
Directional	Occurs when **environmental conditions change**. Selection favours individuals with **extreme phenotypes**.		Mean fur length will increase if the mean temperature decreases in an environment. Individuals with longer fur (i.e. those with more extreme phenotypes) will have higher survival rates. Similarly, mean fur length will decrease if environmental temperature increases.

original population has eight different alleles occurring at various frequencies

chance event reduces the size of the population significantly

the individuals that survive have fewer alleles (just four types) and with different frequencies

as the population recovers, the number and frequency of the alleles are the same as those of the population that came through the bottleneck. This population is less diverse than the original population

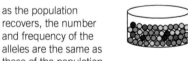

▲ **Figure 1** *An illustration of a genetic bottleneck effect*

Genetic drift

Genetic drift is a random change in allele frequencies. Its effects are more noticeable in small populations. For example, a **genetic bottleneck** is a drastic reduction in population numbers (e.g. due to natural disaster or environmental change). The proportions of alleles in the surviving population could be very different to those in the original population.

The **founder effect** can be considered a type of genetic bottleneck. This occurs when a small group breaks away from the original large population to form a new colony (e.g. a group of birds migrating to a new island).

The Hardy–Weinberg equations

The Hardy–Weinberg equations are used to calculate the proportions of alleles and genotypes in a population. Calculations are carried out for genes with two variants (one dominant and one recessive allele). The first equation we need to understand is:

$$p + q = 1.0$$

Where p = the frequency of the dominant allele, and q = the frequency of the recessive allele. If only two alleles exist, their frequencies must add up to 1.0 (100%).

The second equation considers the frequencies of the possible genotypes in a population:

$$p^2 + 2pq + q^2 = 1.0$$

p^2 = the homozygous dominant genotype; $2pq$ = the heterozygous genotype; q^2 = the homozygous recessive genotype. The frequencies of the three possible genotypes must add up to 1.0 (100%) since no other genotypes are possible.

 ### Worked example of Hardy–Weinberg calculations: Cystic fibrosis

Ireland has the world's highest rate of cystic fibrosis (CF). A recent study estimated that 1 in 1353 births are of babies with CF. From this information alone, how do we calculate the proportion of unaffected carriers of the CF allele in Ireland's population?

1 CF is caused by a recessive allele (f). Only homozygous recessive genotypes produce CF.

2 The homozygous recessive genotype = ff = q^2 in the Hardy–Weinberg equation.

3 q^2 = 1 in 1353 = 0.000 739 098

4 To calculate the frequency of the recessive allele (q) we need to find the square root of q^2. $\sqrt{0.000\,739\,098}$ = 0.027 186 362

5 $p + q = 1.0$, so $p = 1 - q$. We now know $q = 0.027\,186$, therefore $p = 1 - 0.027\,186 = 0.972\,814$.

6 We are now able to calculate the frequency of heterozygous genotypes (Ff) in Ireland. Ff is represented in the Hardy–Weinberg equation by $2pq$. $2pq = 2 \times 0.972\,814 \times 0.027\,186 = 0.052\,894$.

7 A frequency of 0.052 894 means 5.3% of Ireland's population are carriers of the CF allele. Another way of expressing the same value is that approximately 1 in 19 of the population are carriers.

Synoptic link

You learned about evolution and natural selection in Topics 10.4, Evidence for evolution, and 10.8, Changing population characteristics.

Revision tip: Making assumptions

The Hardy–Weinberg equations assume that the following conditions are present in a population: no new mutations, no migration, no natural selection for or against alleles, a large population, and random mating.

In reality, these conditions rarely exist. The Hardy–Weinberg principle nonetheless provides a basis for the study of gene frequencies.

Summary questions

1 Explain how genetic bottlenecks decrease genetic diversity. (*2 marks*)

2 Albinism is a condition in which people lack melanin pigment in their hair, skin, and eyes. It is caused by a recessive allele, which means only homozygous recessive genotypes produce albinism. 1 in 17 000 people worldwide have the condition. Calculate the allele frequency of the recessive allele for albinism. (*2 marks*)

3 Ellis–van Creveld (EVC) syndrome is a disorder that includes symptoms such as dwarfism and extra fingers. The allele that causes the disease is recessive. In one Amish population, 5 in 1000 people have EVC. Calculate the percentage of the Amish population that carries one copy of the EVC allele. 1 in 123 people in the general population carry one EVC allele.
Calculate the difference in the percentage of EVC carriers between the two populations. (*5 marks*)

20.6 Speciation and artificial selection

New species can evolve through natural selection. The process of artificial selection requires human populations to provide a selection pressure by choosing animals and plants with desirable traits.

Speciation

For a new species to evolve, the following events must occur:

- Members of a population become **isolated** from the rest of the population (which prevents gene flow between the two groups).
- Genetic **mutations** continue to occur in both groups, producing **new alleles**.
- The different groups experience different **selection pressures**.
- Different alleles are selected in the two groups.
- Over generations, the two groups become genetically different and **unable to reproduce fertile offspring** together.

Two mechanisms of speciation can occur: **sympatric** and **allopatric** speciation.

Type of speciation	Description	Form of reproductive isolation	Examples
Allopatric	Members of a population are separated by a physical barrier	Geographical	Migration to different islands. A mountain range. Agricultural activity.
Sympatric	Speciation occurs within a population that shares the same habitat. This is very rare, and the theory is controversial.	Temporal Behavioural Mechanical	Differences in the timing of flowering. Different mating rituals/calls. Incompatible reproductive systems.

Artificial selection

By choosing which organisms to breed together over many generations, humans have accelerated the process of evolution. For example, tameness has been selected in cats and dogs, and crop plants have been formed with high yields. This process of artificial selection (**selective breeding**) can cause problems, however, because it requires **inbreeding** (breeding closely related individuals):

- Genetic diversity is reduced.
- Inbred populations are less able to adapt to changing environmental conditions.
- Homozygous recessive disorders are more likely to occur.

> **Revision tip: Defining a species**
> Scientists have long debated the most appropriate definition of a species. A popular definition is 'a group of individuals that can breed to produce fertile offspring'. However, this definition excludes asexual organisms, such as bacteria. An alternative definition (the phylogenetic species concept) is 'a group of organisms sharing similar morphology, genetics, biochemistry, and behaviour'.

Summary questions

1 Suggest the type of reproductive isolating mechanism that exists in the following examples:
   a The Kaibab squirrel and the Abert squirrel are separated by the Colorado river.
   b The white sage plant has a large landing platform for pollinating insects, whereas the black sage has a small landing platform for pollinators.   (*2 marks*)

2 Apple maggot flies lay eggs on either hawthorns or domestic apples. Females tend to lay their eggs on the type of fruit they grew up in, and males usually search for mates on the type of fruit they grew up in. Suggest how speciation might arise in this species.   (*2 marks*)

3 Read the following passage and evaluate the statement 'bonobos and chimpanzees should be considered members of the same species'.   (*2 marks*)
   Bonobos and chimpanzees may have become separated and geographically isolated by the Congo River. However, bonobos and chimpanzees will mate in captivity, and their hybrid offspring are thought to be fertile. Scientists have estimated that more than 1 million years on average is required for reproductive isolation in primates. Chimpanzees and bonobos probably diverged roughly 1 million years ago. The behavioural and anatomical differences between them, which have evolved during their geographical isolation, are insufficient to stop them mating. Nor have post-mating barriers evolved to render any hybrid offspring infertile.

1 Which of the following statements is/are true of the founder effect?

(*1 mark*)

    1 It can be considered a form of genetic bottleneck.

    2 It often results in a change in allele proportions within a population.

    3 It is usually a result of two populations merging.

    A 1, 2, and 3 are correct

    B Only 1 and 2 are correct

    C Only 2 and 3 are correct

    D Only 1 is correct

2 Which of the following statements is/are true of the assumptions made when applying the Hardy–Weinberg principle to calculate the frequency of alleles in a population? (*1 mark*)

    1 Natural selection is operating.

    2 The population is large.

    3 Mating is random.

    A 1, 2, and 3 are correct

    B Only 1 and 2 are correct

    C Only 2 and 3 are correct

    D Only 1 is correct

3 Two parents are both carriers of the recessive allele (f) that causes a rare condition called Friedreich's ataxia (FA).

The parental genotypes are Ff. People with FA have the genotype ff.

The two parents have a daughter. Their daughter, later in life, has a child with a father who is a carrier of FA. What is the probability that this child (the grandchild of the first couple) has FA? (*1 mark*)

    A 0%   C 50%

    B 25%   D 75%

4 Which of the following statements is/are true of recessive epistasis? (*1 mark*)

    1 Two gene loci are involved.

    2 A homozygous recessive genotype is required for the expression of another gene.

    3 AaBb would produce an epistatic effect.

    A 1, 2, and 3 are correct

    B Only 1 and 2 are correct

    C Only 2 and 3 are correct

    D Only 1 is correct

5 Which of the following statements is an example of temporal isolation? (*1 mark*)

    A Separation by a river.

    B Differences in mating displays.

    C Differences in breeding season.

    D Reproductive system incompatibility.

6 Outline, with examples, how the founder effect can change allele frequencies in populations. (*6 marks*)

21.1 DNA profiling

Specification reference: 6.1.3(c), (d), and (e)

A DNA profile is a genetic fingerprint that is unique to each person (except identical twins). DNA profiles are constructed using techniques such as the polymerase chain reaction (**PCR**) and **electrophoresis**.

Producing a DNA profile

To produce a DNA profile, DNA must be:

- **extracted** (and many **copies made** using **PCR**)
- **digested** (broken into fragments) using **restriction endonucleases**
- **separated** using **electrophoresis**
- **hybridised** with **probes** (which bind to fragments and enable them to be visualised)
- **visualised** in banding patterns (bars).

Techniques used in DNA profiling

Technique	Purpose	The process
PCR	DNA amplification (**copying**)	The DNA sample is placed in a thermocycler (which cycles through three temperatures) **95°C**: breaks hydrogen bonds in the DNA, splitting it into two strands **55°C**: primers bond to the end of each DNA strand **72°C**: Taq DNA polymerase joins free nucleotides to each strand.
Electrophoresis	**Separation** of DNA fragments	DNA fragments are placed at the end of a gel plate. A positive electrode is at the opposite end of the plate. DNA moves towards the positive electrode when a current is applied (because all DNA fragments have phosphate groups with negative charges). Longer fragments move slower, shorter fragments move faster. The DNA fragments are therefore separated into bands based on size.

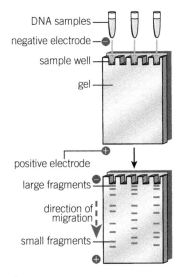

▲ **Figure 1** *Electrophoresis*

Synoptic link

You learned about DNA structure and replication in Topics 3.8 and 3.9.

Maths skill: log scales in PCR

The amplification of DNA is an example of exponential increase. Log scales can be used to show the relationship between cycles of heating and cooling and the increases in copy number.

$$1024 = 2^{10}$$

After 10 cycles, a single fragment of DNA can, in theory, be amplified to 1024 fragments, which would be represented on a log scale as $10^{3.01}$.

Plotting the PCR cycle number against the amount of DNA on a log scale produces a straight line on a graph.

Revision tip: Restriction endonucleases

Restriction endonucleases are enzymes found in bacteria. More than 50 of these enzymes are known, and each one cuts DNA at specific base sequences (recognition sites). As well as digesting DNA prior to electrophoresis, they are used for genetic engineering (see Topic 21.3, Using DNA sequencing).

Revision tip: Taq DNA polymerase

Human DNA polymerase is not used in PCR. Instead, the enzyme is obtained from a thermophilic bacterium, *Thermus aquaticus* (Taq). This form of polymerase is tolerant to heat so does not denature during temperature cycling.

Revision tip: VNTRs

DNA profiles are often formed from sections of DNA called **variable number tandem repeats** (VNTRs). The patterns of VNTRs differ between people; therefore the probability of two unrelated individuals having the same VNTR profile is very low.

Revision tip: Uses of profiling

DNA profiles can be used for:

Forensics (legal applications), such as in criminal investigations and paternity testing.

Disease risk analysis

Classification

Summary questions

1 You learned about chromatography in Topic 3.6, Structure of proteins. Suggest why electrophoresis is sometimes described as 'similar to chromatography'. *(2 marks)*

2 If PCR begins with one fragment, calculate the number of fragments (represented on a log scale) produced after
    a 5 cycles
    b 12 cycles
    c 17 cycles. *(3 marks)*

3 Describe two similarities and two differences between semi-conservative replication of DNA and PCR. *(4 marks)*

21.2 DNA sequencing and analysis
21.3 Using DNA sequencing

Specification reference: 6.1.3(a) and (b)

Scientists are now able to work out the base sequences of genes, which has improved their ability to classify organisms and assess the risk of disease. DNA sequencing also raises the possibility of synthetic biology: the redesign of genes and biological systems.

How is DNA sequenced?

- PCR is conducted (see Topic 21.1, DNA profiling).
- However, some of the free nucleotides in PCR have been modified in two ways:
  - When they bond to a DNA strand they terminate polymerisation.
  - They are fluorescently coloured – A, T, C, and G have different colours.
- New DNA strands stop growing whenever a terminator base is added – PCR is interrupted.
- This results in every possible chain length being produced (e.g. if the full DNA sequence has 1000 bases, chain lengths of 999, 998, 997, etc. will be generated).
- Lasers detect the final base on each chain.
- The sequence of DNA bases can therefore be worked out.

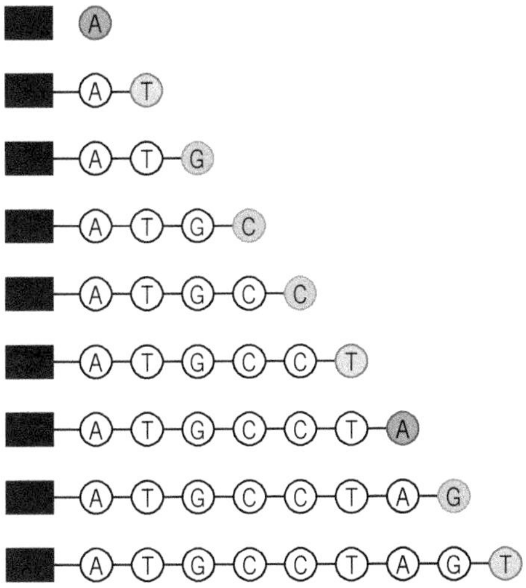

◀ **Figure 1** *DNA is sequenced by producing different chain lengths with terminating bases that are fluorescently tagged*

How are DNA sequences used?

Use	Description
Disease analysis	Particular gene variants can be sequenced and linked to the risk of inheriting certain diseases. Sequencing pathogen genomes enables identification of antibiotic-resistant bacteria, pinpointing genetic markers for vaccines, and identification of targets for drugs.
Classification	Identifying species by using DNA barcodes. Studying evolutionary relationships by comparing similarities and differences between species' base sequences.
Genotype–phenotype relationships	Amino acid sequences do not always match those predicted from base sequences – several phenotypes are possible from the same genotype. Knowledge of both amino acid and base sequences enables comparisons to be made.
Synthetic biology	Genetic engineering (see Topic 21.4) requires knowledge of base sequences.

Common misconception: Bioinformatics

Bioinformatics and *computational biology* are often used interchangeably. However, the terms are subtly different. Bioinformatics is a toolkit – the creation of databases and computer software that can be used to solve biological questions. Computational biology is the application of bioinformatics. For example, the sequencing of genomes relies on bioinformatics and is therefore an example of computational biology.

Go further: Advances in sequencing

The original DNA sequencing method used interrupted PCR in capillaries. Several improvements to the process have been introduced, including:

- The process can now take place on a plastic slide, called a flow cell; millions of DNA fragments can be attached to the slide.

- The use of reversible terminator bases enables a high yield of sequenced DNA.

- The fluorescent tags can be visualised at the same time.

- Pyrosequencing: the binding order of the four bases to a template DNA strand is monitored by the release of light from the luciferase enzyme.

  Many additional sequencing methods are being developed.

1   Suggest what criteria are used to judge the effectiveness of a sequencing method.

Revision tip: Profiles vs. sequences

Do not confuse DNA profiling (producing a genetic fingerprint, unique to a person, which is based on particular sections of DNA) and sequencing (determining the precise base sequences in DNA).

Summary questions

1   Explain the use of sequencing in
    a   the analysis of disease risk.
    b   classification.   (4 marks)

2   Outline the differences and similarities between the use of
    PCR for amplification and PCR for sequencing.   (4 marks)

3   Suggest why protein amino acid sequences sometimes
    differ from those predicted from DNA sequences.   (3 marks)

21.4 Genetic engineering

Specification reference: 6.1.3(f)

Knowledge of base sequences opens up the possibility of altering genes and transferring them between species. This is known as genetic engineering.

Genetic modification techniques

Obtaining the desired gene

A desired gene can be extracted using two methods: producing the gene from an mRNA template, or cutting out the gene using restriction endonucleases.

Enzyme used for gene extraction	Key points
Restriction endonucleases	The gene is cut from its source DNA. Cutting the gene and plasmid with the same enzyme produces two sets of **sticky ends** with complementary base pairings, enabling the gene to be inserted into the plasmid.
Reverse transcriptase	mRNA (transcribed from the desired gene) is extracted from cells. Reverse transcriptase is used to convert mRNA to cDNA (a single strand of complementary DNA).

Using vectors to produce recombinant DNA

A vector is used to transfer the gene into the organism that is being modified. Vectors include:

- **Plasmids**: circular bacterial DNA; by far the most common vector. Bacterial artificial chromosomes (**BAC**s) are synthetic structures based on plasmids. DNA **ligase** is used to seal the gene into the plasmid.

- **Viruses** (e.g. **bacteriophages**, which naturally infect bacterial cells).

The vector, containing the donated gene, must then be transferred into the recipient's cells. This process is called **transformation**. Several methods of transformation are used, as outlined in the table below.

Method of vector transfer	For which organisms is this method used?	Key points
Culture heating	Bacteria	Bacterial cell membranes become more permeable when heated in a calcium-rich solution. Plasmids are then able to diffuse into the cells.
Electroporation	Bacteria and unicellular eukaryotes	Electric current disrupts the cell membrane, enabling plasmids to enter.
Electrofusion	Plants	Electric currents enable the cell and nuclear membranes of two different cells to fuse.
Viral transfer	Plants, bacteria, and animals	Viruses naturally infect cells, and this mechanism can be exploited to insert DNA directly into target cells.
Agrobacterium tumefaciens infection	Plants	*A. tumefaciens* naturally infects plant cells and can be used to introduce recombinant plasmids.

Key terms

Recombinant DNA: DNA combined from two different sources.

Plasmid: Small circular piece of DNA found in prokaryotic cells.

Revision tip: Sticky notes

Many restriction endonucleases produce a staggered cut when digesting DNA. This creates two short sequences of exposed, unpaired bases known as **sticky ends**.

Revision tip: Marker genes

Marker genes (sometimes called reporter genes) indicate whether or not a gene (and therefore a plasmid) has been successfully taken up by bacterial cells. Marker genes are usually for **antibiotic resistance** or **fluorescence**.

Summary questions

1 Explain why, in genetic engineering, it is important to use the same restriction endonuclease to cut both the gene being transferred and the plasmid vector. (*2 marks*)

2 Suggest the advantages of genetically modifying a crop to be herbicide resistant. (*2 marks*)

3 Knockout studies can inactivate a gene by replacing the functional version with a non-functional artificial DNA sequence in a study animal (e.g. mice). Explain how a scientist could use this technology to test the association between a particular gene and the risk of cancer. (*2 marks*)

21.5 Gene technology and ethics

Specification reference: 6.1.3(g) and (h)

Gene technology is a fast-developing field. Some people worry that ethical considerations are being overlooked as the science progresses.

Genetic engineering ethics

Type of engineering	What are the potential ethical issues?
Genetically modified (GM) microorganisms	The use of the technology for biological warfare.
Pest resistance in plants	GM plants could produce toxins that might harm insects other than the targeted pests.
Pharming (producing human medicines from GM animals)	Is animal welfare compromised? Will genetic engineering damage the health of animals?
Patenting (i.e. legal ownership of GM technology)	How available will the technology be for those it might benefit? Companies are able to patent techniques and GM organisms, and farmers in poor countries are forced to pay for their use.

Gene therapy

Gene therapy is the addition of beneficial alleles to the cells of people with disease-causing alleles.

Type of gene therapy	Which cells are targeted?	Examples	Limitations
Somatic cell therapy	Human body cells	Haemophilia, cystic fibrosis, immune diseases	A possible risk of additional health problems. Introduced genes are non-heritable. Requires repeat treatments.
Germline therapy	Gametes or early embryonic cells	None – illegal in humans	Ethical issues – changes are permanent, and who decides which genes are targeted?

Revision tip: Gene therapy transfer methods
Genes can be transferred into target cells using either harmless **viruses** or **liposomes** (hollow spheres of lipid molecules).

Summary questions

1 Why does gene therapy not provide a permanent cure for cystic fibrosis?   (2 marks)

2 Why is gene therapy potentially most useful for treating diseases stemming from single-gene mutations?   (2 marks)

3 Suggest why viruses and liposomes are used as vectors for delivering genes to target cells.   (2 marks)

Chapter 21 Practice questions

1 During the polymerase chain reaction, why is the temperature lowered to 55°C? *(1 mark)*

   A To enable DNA polymerase to function.

   B To enable primers to attach to the DNA strands.

   C To separate DNA strands.

   D To enable free phosphorylated nucleotides to attach to DNA strands.

2 After 12 cycles of PCR, a single fragment of DNA can, in theory, be amplified to how many fragments? *(1 mark)*

   A $10^{3.01}$

   B $10^{3.31}$

   C $10^{3.61}$

   D $10^{3.91}$

3 Which of the following statements is/are true of the sticky ends produced during genetic modification of bacteria? *(1 mark)*

   1 They are often palindromic.

   2 They are produced by DNA ligase.

   3 They are usually more than 20 bases in length.

   A 1, 2, and 3 are correct

   B Only 1 and 2 are correct

   C Only 2 and 3 are correct

   D Only 1 is correct

4 Identify the following molecules that are used in genetic engineering:

   A An enzyme that catalyses the production of new DNA.

   B Several enzymes that can cut DNA at specific sequences.

   C An enzyme that anneals (seals) sticky ends.

   D Circular DNA (often containing antibiotic resistance genes) found in bacteria.

   E An enzyme that converts RNA to DNA. *(5 marks)*

5 Four scientific aims are described below.

   A Generating genetically identical transgenic mice that show disease symptoms.

   B The production of a genetic fingerprint unique to an individual.

   C Transformation using a recombinant plasmid.

   D Genetic manipulation of adult cells to cure a genetic disease.

   Match the correct letters to the following procedures:

   Genetic engineering.

   Somatic cell therapy

   DNA profiling

   Animal reproductive cloning *(4 marks)*

22.1 Natural cloning in plants
22.2 Artificial cloning in plants
Specification reference: 6.2.1(a) and (b)

Many plant species are able to reproduce asexually to form clones. Humans can exploit this cloning potential to produce large numbers of genetically identical plants for commercial benefit.

Natural cloning

Plants form natural clones through a process called **vegetative propagation**. The clones are produced via mitosis from undifferentiated **meristem** cells. Depending on the species, vegetative propagation can be initiated from roots, shoots, or leaves.

Revision tip: How are plants able to clone themselves?

The ability to form natural clones is much more common in plants than animals. This is because many plant cells are undifferentiated and **totipotent** (i.e. retain the potential to divide into a whole organism – see Topic 6.5, Stem cells). Totipotent **stem cells** are located in regions of plants called **meristem** tissue (found principally in roots, shoots, and cambium).

▼ **Table 1** *Examples of vegetative propagation forming natural clones in plants*

Type of vegetative propagation	Example	Description
Root suckers	Elm trees	Root suckers/basal sprouts grow from meristem cells in the tree trunk close to the ground.
Runners	Strawberry plants	Stems grow sideways along the ground from the parent plant. Roots develop where the runner touches the ground.
Tubers	Potatoes	Underground stems swell with nutrients and develop into new plants.
Bulbs	Daffodils	Leaf bases swell with nutrients, and buds develop into new plants.
Rhizomes	Marram grass	Underground stems develop buds and form new vertical shoots.

Practical skill: Plant cuttings

Humans can produce clones of plants by cutting and replanting stems. This exploits and speeds up plants' natural cloning processes. The key aspects of growing cuttings are:

- Short sections of stems are cut and planted.

- Rooting hormone is applied to encourage root growth.

- The cutting is watered thoroughly and covered with a plastic bag for several days to create warm, moist conditions.

Artificial cloning

Humans can produce many genetically identical plants through **micropropagation**. This is especially useful when the desired plant is rare, produces few seeds, or does not readily produce natural clones.

Practical skill: Micropropagation

The key aspects of micropropagation are:

- A small sample of meristem tissue is taken from shoot tips. The tissue that is removed is called an **explant**.

- The explant is **sterilised** (e.g. in ethanol or sodium dichloroisocyanurate).

- The sterilised explant is placed in a **culture** medium (i.e. a solution containing an ideal balance of **nutrients** and plant **hormones**).

- The explant cells divide to form a mass of undifferentiated cells called a **callus**.

- The callus is transferred to a new culture medium, which contains hormones that encourage differentiation and shoot growth.

- The developing plantlets are transferred into soil.

Synoptic link

You learned about the roles of various plant hormones in Chapter 16, Plant responses.

Pros and cons of artificial cloning

▼ **Table 2** *Advantages and disadvantages of artificial cloning by micropropagation*

Advantages	Disadvantages
Rapid production. **Seedless**, sterile crop plants (e.g. seedless grapes) can be produced. The genetic make-up of the propagated plants is known and **desired traits** can be retained in the clones.	Quite **expensive**. The **lack of genetic variation** means all the clones are vulnerable to the same diseases or environmental change.

Summary questions

1 Describe how the presence of meristem tissue in many plant species enables them to form natural clones. *(3 marks)*

2 Plant hormones are added to cultures during micropropagation. A balanced ratio of auxin to cytokinins promotes callus formation. Suggest why auxin concentration might be raised above cytokinin concentration at certain stages of micropropagation. *(2 marks)*

3 Explain why vegetative propagation is likely to be more effective at enabling plants to recover from a fire than a pathogenic infection. *(2 marks)*

Key terms

Vegetative propagation: Asexual reproduction in plants to produce clones.

Micropropagation: The use of tissue culture to produce many artificial clones of a plant.

Clones: Genetically identical individuals that result from asexual reproduction.

22.3 Cloning in animals

Adults of some animal species can produce clones, although this is less common than cloning in plants. Scientists have developed techniques to produce clones of commercially beneficial animals.

Natural cloning

Many invertebrate animals produce clones (e.g. species of *Hydra* generate buds that develop into clones, and starfish form from fragments of an original animal). Very few vertebrates are known to produce clones of themselves. The endangered sawfish is the first vertebrate species to be observed successfully cloning in the wild.

> **Revision tip: For the twin!**
> **Monozygotic twins** are produced when an embryo splits at an early stage of development to form two separate embryos. The two resultant offspring originate from the same zygote and are clones of each other (but are not clones of either parent).

 Go further: Aphids

Aphids (sometimes called greenflies) are insects. Some species of aphid can reproduce both sexually and asexually, depending on the conditions in their environment. During spring and summer, asexual reproduction generates clones of female aphids; this process is called **parthenogenesis** and offspring are born as nymphs without the need for eggs. During the autumn and early winter, sexual reproduction produces eggs that hatch the following spring.

1 Suggest what benefit aphids gain from using asexual reproduction during the summer.

2 Suggest why aphids switch to sexual reproduction prior to the winter.

Artificial cloning

Scientists can now produce clones of vertebrates using two techniques: **embryo splitting** (which mimics the natural twinning process) and **somatic cell nuclear transfer**.

▼ **Table 1** *The two methods for artificially cloning animals*

	Embryo splitting/artificial twinning	Somatic cell nuclear transfer (SCNT)
What is done?	Sperm from a male with desired traits is used to fertilise eggs from a desired female (via **artificial insemination** or *in vitro* fertilisation). An **embryo is split** into several smaller embryos and grown in a lab. Each embryo is implanted into a **surrogate mother**.	The **nucleus** from an adult somatic (body) cell is transferred into an **enucleated egg** (an egg lacking its own nucleus) using **electrofusion**. The resultant embryo is transferred into a **surrogate mother**.
Nature of the clones	Offspring are clones of each other, but (like regular offspring) share 50% of their alleles with the father and 50% with the mother.	Offspring are clones of the original body cell (from which the nucleus was taken).

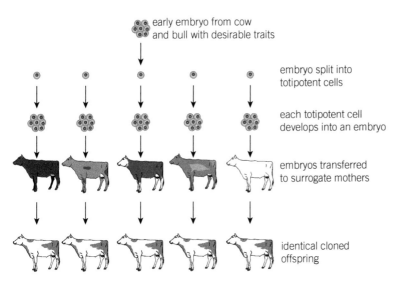

▲ **Figure 1** *Artificial cloning by embryo splitting*

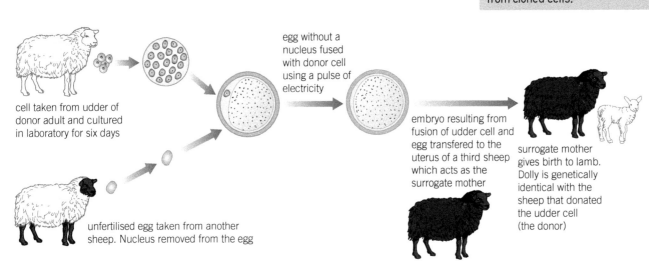

▲ **Figure 2** *Artificial cloning by somatic cell nuclear transfer*

▼ **Table 2** *The pros and cons of artificial animal cloning*

Pros of animal cloning	Cons of animal cloning
Embryo splitting enables **many more offspring** to be produced from the best farm animals. SCNT has the potential to reproduce specific animals, such as pets, and rare or extinct animals.	SCNT is very **inefficient** (i.e. high failure rate). Cloned animals may have shortened lifespans or **health problems**.

Summary questions

1 Describe the difference between reproductive and non-reproductive cloning. *(2 marks)*

2 Explain why an embryo should be split at an early stage for artificial twinning to succeed. *(2 marks)*

3 A student suggested that clones produced from somatic cell nuclear transfer would contain some DNA from the egg donor. Is the student correct? *(2 marks)*

22.4 Microorganisms and biotechnology
22.5 Microorganisms, medicines, and bioremediation

Specification reference: 6.2.1(e) and (f)

Key term

Biotechnology: The exploitation of organisms and biological processes in industry, food science, agriculture, or medical science.

Scientists can exploit the reactions carried out by microorganisms for a variety of commercial and industrial processes.

Uses of microorganisms

▼ **Table 1** *Some of the uses of microorganisms in biotechnology*

Process	Examples	Microorganism used
Food production	Brewing (**alcohol** production) – anaerobic respiration by the fungus produces ethanol.	Brewer's yeast (a fungus)
	Baking (**bread** production) – the CO_2 produced by the fungus makes bread rise.	Baker's yeast (a fungus)
	Cheese and yoghurt production.	*Lactobacillus* bacteria
	Mycoprotein production (which can be eaten as a meat substitute).	*Fusarium* fungus
	Fruit juice – pectinase breaks down pectin in fruit and releases juice.	Pectinase enzyme from *A. niger* fungus
Drug manufacture	Penicillin **antibiotic**. (Antibiotics are secondary metabolites, which you will learn about in Topic 22.6, Culturing microorganisms in the laboratory.)	*Penicillium* fungus
	Insulin.	Genetically modifed *E. coli* bacteria
Bioremediation	**Water treatment**.	Various bacteria and fungi

Synoptic link

E.coli can be genetically modified to produce insulin. You learned about the principles of genetic modification in Topic 21.4, Genetic engineering.

Which characteristics make microorganisms useful?

▼ **Table 2** *The characteristics that make microorganisms useful in biotechnology*

Benefit of using microorganisms	Detail
Fewer ethical issues	The welfare issues associated with the use of animals are not present.
Short life cycle	Microorganisms reproduce rapidly; large numbers can be produced in a short period of time.
Genetic engineering	Bacteria are relatively easy to genetically engineer.
Simple nutrition	Nutrient requirements are cheap; they can often be grown on waste materials.
Fewer energy requirements	Most microorganisms require only low temperatures.

Revision tip: Mycoprotein – food for thought?

Mycoprotein (produced by *Fusarium* fungi) can be used as an alternative to meats. Other than the benefits listed in Table 2, mycoprotein is useful because of its low fat content and our ability to manipulate the final taste of the product. However, some people are disturbed by the thought of eating products from microorganisms.

Summary questions

1 *Aspergillus* fungi can ferment soya beans, and this process is used to produce soya sauce. Explain why this is an example of biotechnology.
(1 mark)

2 Suggest why microorganisms are used to treat waste water (i.e. bioremediation).
(2 marks)

3 Suggest what advantage *Penicillium* may gain by producing antibiotic chemicals such as penicillin.
(2 marks)

22.6 Culturing microorganisms in the laboratory
22.7 Culturing microorganisms on an industrial scale

Specification reference: 6.2.1(g) and (h)

Scientists grow populations of microorganisms in the laboratory or on an industrial scale by culturing them (i.e. providing a nutrient medium that encourages their growth and replication).

Bacterial colony growth curve

The growth of bacterial populations follows a standard pattern and can be divided into four phases.

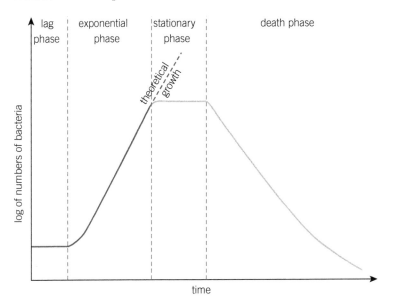

▲ **Figure 1** *A typical growth curve of bacterial populations*

▼ **Table 1** *The phases of a growth curve*

Phase	What happens?
Lag	Genes for important enzymes are transcribed, and bacteria adapt to their new environment.
Exponential (log)	The rate of reproduction is close to maximum and the population size increases at an exponential rate.
Stationary	The population reaches its maximum size (note: this could be referred to as *carrying capacity* in a natural environment); death rate = reproductive rate.
Death (decline)	Nutrients are exhausted, waste products are produced, and death rate rises.

Several **limiting factors** stop exponential growth, including: nutrient levels, oxygen availability, temperature, waste products, and pH.

Metabolite production

Metabolites are products of an organism's metabolism. **Primary metabolites** are substances formed as part of the normal growth of microorganisms (e.g. proteins, enzymes, and ethanol). Their rate of production follows the standard growth curve of bacteria. **Secondary metabolites** (e.g. antibiotic chemicals) are produced principally in the stationary phase. Not all microorganisms produce secondary metabolites.

Synoptic link

You have encountered the idea of limiting factors in Topic 17.4, Factors affecting photosynthesis, and you will read about them again in Topic 24.1, Population size.

Synoptic link

You learned how serial dilutions are carried out in Topic 16.1, Plant hormones and growth in plants.

Summary questions

1   Describe three factors that may result in the initiation of a death phase in a bacterial population. *(3 marks)*

2   Explain why the conditions in a *Penicillium* batch culture encourage the production of penicillin. *(3 marks)*

3   Bacterial cells were cultured in a 20 cm³ nutrient broth. 1 cm³ of the broth was placed in 9 cm³ of water to form a new solution (B). 1 cm³ of solution B was transferred to 9 cm³ of water to form solution C. 0.1 cm³ of solution C was placed on one agar plate and the same volume was transferred to another agar plate. Sixteen colonies grow on one plate, and twelve grew on the other plate. Estimate the population size in the original nutrient broth. Express your answer in standard form. *(4 marks)*

Laboratory cultures

The nutrient medium in which microorganisms are cultured can be liquid (broth) or solid (agar). Cultures need to contain the correct nutrients, be kept at the correct temperature, and be sterile. **Aseptic techniques** are used to keep the nutrient media sterile (e.g. preventing contamination of cultures from the air; using a sterilised inoculating loop to transfer bacteria to agar).

Worked example: serial dilutions

The number of cells in a population of microorganisms can be estimated by using serial dilutions. For example, imagine that we have a culture of bacteria in 10 cm³ of nutrient broth.

- 1 cm³ of the broth is transferred to 9 cm³ of water. This is a 10^{-1} dilution, and the new solution should contain 10% of the original cells.

- 1 cm³ of the new solution is transferred to 9 cm³ of water. This third solution will now contain 1% of the original cells.

- 0.1 cm³ of the final solution is then transferred to an agar plate. The agar plate should contain 0.01% of the original cells.
  10 cm³ / 0.1 cm³ = 100, and 1% / 100 = 0.01%.

- 40 colonies develop on the agar plate. We can assume each colony has grown from a single bacterial cell.

- The number of cells in the original broth = (100% / 0.01% = 10 000) × 40 = 400 000.

Industrial cultures

On an industrial scale, microorganisms are cultured in reaction vessels called bioreactors.

▼ **Table 2** *Which factors are controlled in industrial cultures?*

What aspect is controlled?	Why is this controlled?
Temperature	To maintain the optimum temperature for enzymes.
Nutrients and oxygen	The medium is mixed and stirred to ensure an even distribution of respiratory substrates and oxygen. Levels can be monitored and extra nutrients and oxygen are added if necessary.
Asepsis	To prevent contamination from other, unwanted microorganisms.

Industrial cultures can be run using either **batch** or **continuous** culture.

▼ **Table 3** *A comparison of batch and continuous cultures*

	Batch	Continuous
How much nutrient medium is used?	A fixed volume of nutrient medium is used	Nutrient medium is added during the process
How long is the culture in the bioreactor?	A fixed time period	Indefinitely
What happens to the population size?	Waste and population builds up (but the process is halted before the death phase)	Microorganism population and waste are removed continuously to maintain population size
What can be produced?	Secondary metabolites (e.g. antibiotics)	Primary metabolites (and GM products such as insulin)

22.8 Using immobilised enzymes

Specification reference: 6.2.1(i)

As you learned in Topic 22.4, Microorganisms and biotechnology, enzymes from other organisms have been used by humans for many years. Rather than using the whole organism, enzymes can be isolated; this is more efficient. However, to improve efficiency even further, enzymes tend to be immobilised rather than free in solution.

Using immobilised enzymes

Synoptic link

You learned about enzymes in Chapter 4, Enzymes. Types of bonding (e.g. ionic and covalent) were discussed in Topic 3.6, Structure of proteins.

▼ **Table 1** *Advantages and disadvantages of using immobilised enzymes*

Advantages of using immobilised enzymes	Disadvantages of using immobilised enzymes
Less downstream processing (i.e. once the reaction has occurred, enzymes do not need to be separated from the product). Enzymes can be immediately **reused** (saving money). Enzymes can be more protected and therefore **more stable** and reliable.	**More expensive** and time-consuming to **set up**. Sometimes **less active** and therefore less efficient than freely dissolved enzymes.

How can enzymes be immobilised?

▼ **Table 2** *Methods of immobilising enzymes*

Method		What is done?	Pros	Cons
Surface immobilisation	Adsorption	Enzymes are attached to inert material (e.g. glass or alginate beads).	Relatively cheap.	Weak attachment increases the risk of enzymes leaking. Relatively slow.
	Covalent and ionic bonding	Enzymes form covalent bonds with silica gel or clay particles, or cross-linked to each other. Ionic bonds can form with cellulose or synthetic polymers.	Little enzyme leakage. The enzymes' active sites are very accessible for substrates.	Cost varies.
Entrapment	In matrix	Enzymes are trapped in a gelatin or cellulose matrix.	Active sites are not altered by the immobilisation.	Expensive. Active sites are less accessible for substrates. Diffusion and collection of the product can be slow.
	Membrane entrapment	Separation from the substrate solution using a semi-permeable membrane.		

Examples of immobilised enzyme use

Immobilised enzymes are especially useful for generating large quantities of product because continuous production is possible.

▼ **Table 3** *Examples of immobilised enzymes being used in biotechnology*

Enzyme	Product
Glucoamylase	Glucose (formed by breaking down dextrins)
Lactase	Glucose and galactose; lactose is broken down to produce lactose-free milk
Aminoacylase	L-amino acids (for drugs and cosmetics)
Penicillin acylase	Semi-synthetic penicillin (antibiotic)
Glucose isomerase	Fructose (which is used as a sweetener in food and drink)

Revision tip: Extra special

Extracellular enzymes are more commonly used than intracellular enzymes in research and for commercial processes. This is because extracellular enzymes are easier to isolate and are usually adapted to withstand a greater range of temperatures and pH than intracellular enzymes.

Summary questions

1 Describe how enzymes can be immobilised by entrapment. *(2 marks)*

2 Evaluate the economic advantages and disadvantages of using immobilised enzymes. *(3 marks)*

3 Immobilised penicillin acylase is used to produce semi-synthetic penicillin. Suggest why this may be very useful for the treatment of bacterial diseases. *(2 marks)*

1 Complete the following passage about natural cloning in plants by selecting the most appropriate words or phrases to place in the gaps.

(4 marks)

Many plants can produce clones naturally in a process known as vegetative .... The clones develop from undifferentiated tissue called .... For example, elm trees produce .... from their roots, which develop into .... identical clones of the original tree.

2 The following events form part of the somatic cell nuclear transfer procedure for producing artificial animal clones. Place the events in the correct chronological order. *(4 marks)*

   A The embryo is placed in a surrogate mother.

   B A nucleus is removed from one of the donor cells.

   C Electrofusion.

   D A cell is taken from an adult donor and cultured for several days.

3 Explain the changes in glucose and penicillin concentration shown in the graph.

(4 marks)

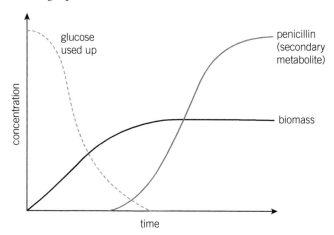

4 Suggest the form of enzyme immobilisation represented by each of the four diagrams below. *(4 marks)*

 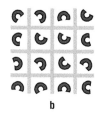 

a   b   c   d

23.1 Ecosystems
23.2 Biomass transfer through an ecosystem

Specification reference: 6.3.1 (a) and (b)

An ecosystem consists of organisms and the non-living (abiotic) factors with which they interact. As organisms consume other organisms, chemical energy (stored as biomass) is passed through an ecosystem. These energy transfers, however, are far from 100% efficient.

Factors affecting ecosystems

▼ **Table 1** *Examples of biotic and abiotic factors*

Factor	Examples	How does it affect organisms?
Biotic	Living organisms	Competition and consumption
Abiotic	Temperature	Enzyme activity Thermoregulation Leaf-fall and flowering (in plants)
	Light	Photosynthetic rate (in plants)

Food chains and trophic levels

A food chain illustrates the transfers of energy between organisms within an ecosystem. Each stage in a food chain is called a **trophic level**. Food chains vary in length. Two examples are shown below.

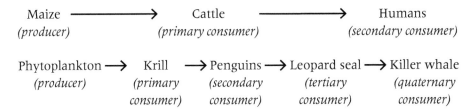

Maize ⟶ Cattle ⟶ Humans
(producer)   *(primary consumer)*   *(secondary consumer)*

Phytoplankton → Krill → Penguins → Leopard seal → Killer whale
(producer)   *(primary consumer)*   *(secondary consumer)*   *(tertiary consumer)*   *(quaternary consumer)*

Measuring biomass transfers between trophic levels

Transfers between trophic levels can be measured as biomass or energy (because biomass consists of stored chemical energy).

Biomass can be calculated as either wet or dry mass.

▼ **Table 2** *How biomass can be measured*

Measure of biomass	Procedure	Advantage	Disadvantage
Fresh (wet) mass	Living organisms measured.	No organisms are killed.	The presence of water reduces accuracy.
Dry mass	Organisms are killed and heated to 80°C until all water has been removed.	The estimates of mass are more accurate.	Organisms must be removed from the ecosystem and killed.

The trophic levels in a food chain are shown graphically in **pyramids**. The steps of the pyramid can represent numbers, biomass, or energy.

Key terms

Ecosystem: All of the biotic and abiotic factors in an area and their interactions.

Trophic level: A stage a food chain.

Producer: An organism (e.g. plant) that converts light energy to chemical energy (i.e. autotrophic nutrition).

Consumer: An organism that gains energy by feeding on other organisms (i.e. heterotrophic nutrition).

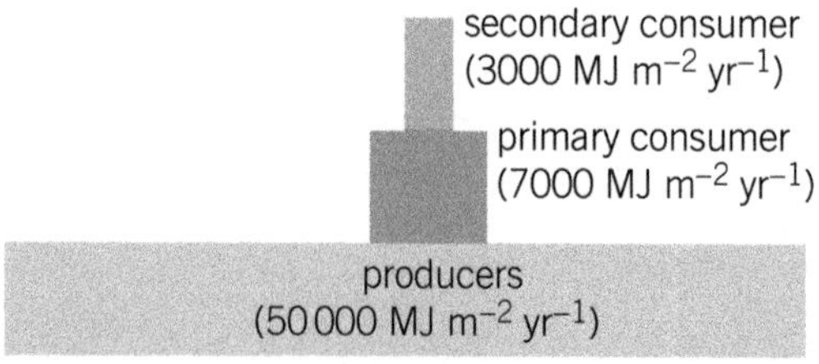

secondary consumer
$(3000\ MJ\ m^{-2}\ yr^{-1})$

primary consumer
$(7000\ MJ\ m^{-2}\ yr^{-1})$

producers
$(50\,000\ MJ\ m^{-2}\ yr^{-1})$

▲ **Figure 1** *A pyramid of energy*

Revision tip: When is a pyramid not a pyramid?

Pyramids of energy and biomass, if measured over a full year, will always be pyramid-shaped. Energy and biomass values decrease from producers up through the consumer levels. However, sometimes a pyramid of numbers will have a relatively narrow bar at the producer level. For example, trees may be few in number within an ecosystem, but their total biomass will still be greater than that of the primary consumers.

Maths skill: Units for energy and biomass

Biomass is usually measured in $g\,m^{-2}\,yr^{-1}$ (grams per square metre per year) on land or $g\,m^{-3}\,yr^{-1}$ (grams per cubic metre per year) in water.

Energy is usually measured in $kJ\,m^{-2}\,yr^{-1}$ (or $kJ\,m^{-3}\,yr^{-1}$).

The units include a standard area (m^2) to enable comparisons to be made between ecosystems of different sizes.

You may be asked to convert between units. For example, $300\,000\ kJ\,m^{-2}\,yr^{-1}$ = $300\ MJ\,m^{-2}\,yr^{-1}$ (because $1\ MJ = 1000\ kJ$) = $300\,000\,000\ MJ\,km^{-2}\,yr^{-1}$ (because there are one million square metres in one square kilometre) = $3 \times 10^8\ MJ\,km^{-2}\,yr^{-1}$ (in standard form).

Efficiency of transfers

Not all the energy in one trophic level is transferred to the next level. Some energy is released as heat, and some stored chemical energy (in biomass) cannot be consumed. The percentage of energy that is transferred represents the efficiency of the transfer.

▼ **Table 3** *The efficiency of energy transfers to producers and consumers*

	Efficiency at producer level	Efficiency at consumer levels
What transfer occurs?	Light energy is converted to chemical energy in producers.	The stored energy in the biomass of one trophic level is transferred to the next trophic level by consumption.
Why is some energy not transferred?	Most light energy (90%) cannot be absorbed by plants. Some absorbed energy is used in the reactions of respiration and is not converted to chemical energy in biomass.	Some parts of organisms (e.g. bones, roots, feathers) are inedible or indigestible. Some energy is lost from the food chain as heat, through movement, or in urine.
Calculation formula	Net production = gross production − respiratory losses	Ecological efficiency = (energy available after the transfer / energy available before the transfer) × 100

Worked example: Efficiency calculations

The net production in a grassland ecosystem was measured as $45\,g\,m^{-2}\,year^{-1}$. In one year, a buffalo consumes the equivalent of all the biomass in a 350 m × 350 m area. 600 kg of grass is converted into biomass in the buffalo.

Calculate the efficiency of the energy transfer between the grass and the buffalo.

Step 1: Calculate the total biomass consumed by the buffalo.
$45 \times (350 \times 350) = 5\,512\,500\,g = 5512.5\,kg$.

Step 2: Calculate the efficiency of the transfer. $(600\,kg\,/\,5512.5\,kg) \times 100 = 10.9\%$

How can humans manipulate biomass transfers?

Humans can tailor their farming practices to maximise production in plants and the efficiency of energy transfer to primary consumers.

▼ **Table 4** *How farming practices improve the efficiency of energy transfers*

	Which factor is manipulated?	How can humans manipulate the factor?
Production in plants	Light	Plants are grown in greenhouses under optimal light intensity and duration. Seed sowing is timed to maximise the leaf area present for photosynthesis during the brightest months of the year.
	Temperature	Greenhouses provide regulated, optimal temperatures.
	Water	Irrigation. Genetically engineered drought resistant crops.
	Nutrient levels	Fertiliser use.
	Pests	Pesticide use.
Efficiency of energy transfers to primary consumers	Movement	The movement of farm animals is limited. More energy is channelled into growth.
	Disease	Antibiotic use reduces energy expenditure in immune systems.

Summary questions

1 Describe the factors that reduce the efficiency of energy transfer between trophic levels. *(3 marks)*

2 Explain how farmers can increase the proportion of consumed energy that is used for growth in cattle. *(3 marks)*

3 Over a 2 year period, a farmer grew 200 000 kg of a crop in a 1 km² field.
  a Calculate the net primary production of this crop $(kg\,m^{-2}\,yr^{-1})$. *(2 marks)*
  b Explain how genetic engineering could increase the primary production of the crop. *(3 marks)*

23.3 Recycling within ecosystems

Specification reference: 6.3.1 (c)

Energy flows through an ecosystem and is transferred into the atmosphere. Nitrogen and carbon, however, are recycled back into ecosystems from the atmosphere.

Decomposition

Decomposition is the process of large organic molecules (in dead animals and plants) being broken down into smaller inorganic molecules. Decomposers can be either **bacteria** or **fungi**; they are also referred to as **saprotrophs** (i.e. they feed on dead or waste organic matter).

Decomposition in the nitrogen cycle is called **ammonification** (the breakdown of proteins, nucleic acids, and vitamins in dead organisms, faeces, and urine to form ammonia (NH_3) and ammonium (NH_4^+) compounds).

The nitrogen cycle

Nitrogen passes through a food chain as consumers feed on organisms in lower trophic levels. Decomposition temporarily removes nitrogen from a food chain. Three processes (outlined in the table) recycle nitrogen so that it eventually re-enters the food chain.

▼ **Table 1** *Reactions in the nitrogen cycle*

Process	What is the reaction?	Which bacteria carry out the reaction?
Nitrification	NH_3/NH_4^+ ions to NO_2^- (**nitrite** ions)	*Nitrosomonas*
	NO_2^- (nitrite ions) to NO_3^- (**nitrate** ions)	*Nitrobacter*
Nitrogen fixation	N_2 (in the atmosphere) is converted to NH_4^+ ions/NH_3	*Rhizobium* (mutualistic – lives in the roots of some plants)
		Azotobacter (free-living, in the soil)
Denitrification	NO_3^- to N_2 gas	Denitrifying bacteria (which require anaerobic conditions)

Revision tip: Oxidation or reduction?

You can also think about the processes in the nitrogen cycle in terms of chemistry – i.e. is nitrogen being oxidised or reduced? **Nitrification** represents **oxidation. Nitrogen fixation** and **denitrification** are examples of **reduction** reactions.

Revision tip: An absorbing choice?

Many plants tend to absorb nitrates (NO_3^- ions) from the soil. However, some plants can absorb ammonia (NH_3) /ammonium ions (NH_4^+) instead. This is especially true in soils with anaerobic conditions, where nitrification is difficult.

Remember, NH_3 is produced inside plant roots in species that contain *Rhizobium*, so it would be incorrect to write that these plants absorb NH_3 from soils.

Revision tip: Nitrogen and its compounds

It is easy to muddle the terms you will be using when describing the nitrogen cycle. Remember that 'nitrogen' is the element N, which is found in the atmosphere as a gaseous molecule, N_2.

Compounds containing nitrate ions (NO_3^-) or ammonium ions (NH_4^+) can be described as 'nitrogen-containing compounds' (e.g. potassium nitrate (KNO_3) and ammonium sulfate (NH_4SO_4)), as can ammonia (NH_3). If you are in doubt about any chemical formula you should use the full name.

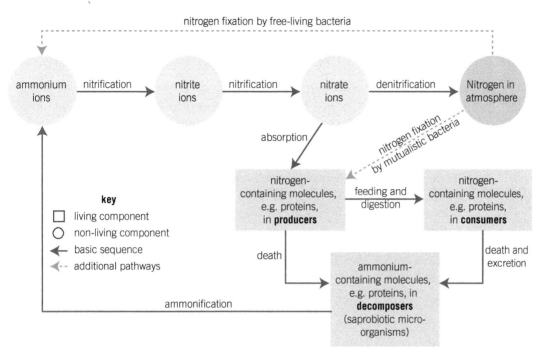

▲ **Figure 1** *The nitrogen cycle*

The carbon cycle

Photosynthesis converts CO_2 to organic molecules in organisms. **Respiration**, **decomposition**, and **combustion** return CO_2 to the atmosphere. These processes are naturally in balance, which maintains relatively constant atmospheric CO_2 concentrations. However, human activity (e.g. more combustion and deforestation) has increased the rate of CO_2 production.

> ### Revision tip: Fuelling up
> Fossil fuels form when aerobic decomposition is prevented (i.e. in anaerobic conditions).

> ### Synoptic link
> Some of the features of fungi and bacteria were covered in Topic 10.2, The five kingdoms. You learned about photosynthesis in Chapter 17, Energy for biological processes, and respiration in Chapter 18, Respiration.

Summary questions

1 Add the most appropriate words to complete this paragraph, which describes the nitrogen cycle.
Ammonium ions can be added to the soil through two processes, _____ (controlled by bacterial species such as *Azotobacter*) and _____ (controlled by saprotrophic bacteria). Nitrifying bacteria convert ammonium ions to _____ ions and nitrate ions. In anaerobic conditions, nitrate ions are converted to nitrogen gas in a process known as _____ . *(4 marks)*

2 Explain how human activity has upset the natural balance of the carbon cycle. *(3 marks)*

3 Suggest why farmers may try to prevent their soils becoming waterlogged. *(3 marks)*

23.4 Succession

The composition of an ecosystem gradually alters as the abiotic factors in the environment change. The progression from bare ground to a stable, complex community of organisms is known as **succession**.

Primary and secondary succession

Primary succession begins with bare ground (lacking soil) that has been newly formed (e.g. bare volcanic rock that becomes exposed). The starting point for secondary succession is bare soil that has resulted from deforestation or fire.

> **Question and model answer: Primary succession**
>
> Q. Describe the process of primary succession from bare rock to a stable community.
>
> A.
>
> - Only **pioneer species** can **colonise** the bare rock. These are species that are adapted to extreme conditions (e.g. exposure to wind and high light intensity) and can fix nitrogen from the atmosphere. Examples include lichen and algae. Pioneer species represent the first **seral stage**.
>
> - The erosion of the rock produces a basic soil.
>
> - The death and decomposition of pioneer organisms adds nutrients to the soil.
>
> - Soil development enables other species to colonise (i.e. secondary colonisers such as mosses).
>
> - Improved environmental conditions enable more species to colonise (which outcompete the pioneers and secondary colonisers).
>
> - Eventually a **climax community** forms. This is stable (i.e. if environmental conditions remain the same, the species numbers will not vary much), and relatively biodiverse.

Deflected succession

Human activities can prevent a climax community from forming. This is known as **deflected succession** and results in the formation of a community known as a **plagioclimax**. Two examples of plagioclimax communities are found on agricultural land and in managed forests.

Revision tip: What about animals?

Succession also occurs among animal communities. As the composition of plant species in a community develops, this enables different animal species to enter the ecosystem and exploit new niches.

Summary questions

1. Explain what is meant by a plagioclimax, and identify one example resulting from land management. *(2 marks)*

2. Describe three important characteristics of a pioneer species. *(3 marks)*

3. Describe the difference between primary succession and deflected succession, and explain why primary succession is likely to result in greater biodiversity. *(3 marks)*

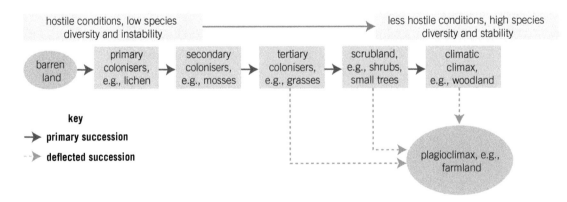

▲ **Figure 1** *Ecological succession*

23.5 Measuring the distribution and abundance of organisms

Specification reference: 6.3.1 (e)

You learned about sampling techniques and strategies in Chapter 11, Biodiversity. Here you will examine how these methods are employed to measure the distribution and abundance of organisms.

Measuring distribution

A **transect** is used to analyse how the distribution of organisms varies in an ecosystem. This is a form of **systematic** sampling (see Topic 11.2, Types of sampling).

Measuring abundance

Sampling allows population sizes to be estimated. **Quadrats** can be used to estimate the abundance of a plant species per area (see Topic 11.3, Sampling techniques).

 Worked example: Estimating plant abundance

A 1 m² quadrat was used to take 8 samples in a 200 m² field. 70 dandelion plants were counted. Estimate the dandelion population size in the field.

Step 1: Individuals m^{-2}
= Number of individuals counted / total area sampled $(m^2) = 70/8 = 8.75\,m^{-2}$

Step 2: Multiply the calculated value by the total area of the habitat. i.e.
$8.75 \times 200 = 1750$

Animal abundance is estimated using the **capture-mark-release-recapture** technique.

 Worked example: Estimating animal abundance

Scientists captured and marked 13 water voles along a river ecosystem. One week later they captured 14, of which 2 were marked.

Estimate the total vole population in the river.

Step 1: Choose the correct equation, which is:
Number of individuals in the first sample × number in the second sample / number of marked individuals in the second sample.

Step 2: Enter the measured values into the equation: $13 \times 14 / 2 = 91$

The total vole population is estimated to be 91.

Summary questions

1 A 1 m² quadrat was used to take 500 samples in a 1 km² grassland. 48 blue fleabane plants were counted. Estimate the blue fleabane population size in the grassland. *(2 marks)*

2 Explain why the measurement of distribution should be systematic. *(2 marks)*

3 A student captured and marked 8 slugs in a field. One week later the student captured 16, of which 3 were marked. Estimate the total slug population in the field. *(2 marks)*

Revision tip: Which transect?

Two forms of transect exist. A **line transect** involves taking samples at regular intervals along a linear tape. It will generally show distribution but give no information about abundance. A **belt transect** involves recording species within two parallel lines. It provides both distribution and abundance data.

Revision tip: Calculating biodiversity

Once abundance data have been collected for all the species in a habitat, biodiversity can be calculated using Simpson's Index of Diversity (see Topic 11.4, Calculating biodiversity).

Synoptic link

You learned the fundamentals of ecological sampling in Topics 11.2, Types of sampling, and 11.3, Sampling techniques. Topic 11.4, Calculating biodiversity, covered how to use the Simpson's Index of Diversity equation.

1 A farmer grew 80 000 kg of a crop in a 480 m² field over a period of one year. What is the net primary production of this crop (units = g m⁻² yr⁻¹)? *(1 mark)*

  A 1.67×10^2 C 1.67×10^4

  B 1.67×10^3 D 1.67×10^5

2

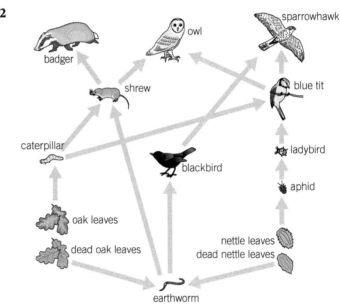

The diagram shows part of a woodland food web.

a Name all of the producers shown in the food web. *(1 mark)*

b Name all of the tertiary consumers shown in the food web. *(1 mark)*

c Suggest the short-term effects on the populations of other species if the ladybird population were to disappear from this ecosystem. *(4 marks)*

d Suggest which energy transfer in this food web is likely to be the least efficient. Explain your answer. *(2 marks)*

3 Use to the diagram on the left to calculate the ecological efficiency of energy transfer between:

a smelt and humans *(1 mark)*

b trout and humans *(1 mark)*

c algae and small aquatic animals. *(1 mark)*

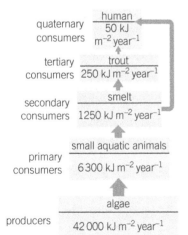

4 Match the following sentences to the correct process in the nitrogen cycle. *(3 marks)*

A Carried out by *Nitrobacter*.

B Carried out by *Nitrosomonas*.

C Produces nitrogen gas.

D Produces nitrate ions.

E Produces ammonia.

F Removes nitrogen gas.

G Removes nitrate ions.

Processes:

Nitrification   Denitrification   Nitrogen fixation

24.1 Population size

Specification reference: 6.3.2 (a)

The size of a population is dictated by many biotic and abiotic factors, some of which limit the number of individuals the population can support. The maximum size of a population is known as its **carrying capacity**.

Population growth curves

Growth curves tend to be sigmoidal in shape. At first, populations grow slowly before experiencing a period of rapid growth. Eventually, population size plateaus when the carrying capacity is reached; numbers will fluctuate during this stable state but remain relatively constant.

Factors that determine population size

Limiting factors determine the size to which a population can grow.

▼ **Table 1** *Density-dependent and density-independent factors*

	Definition	Examples
Density-dependent factors	The impacts of these factors vary with population density.	**Abiotic**: temperature, water availability, pH, light intensity. **Biotic**: predation, disease, competition, migration.
Density-independent factors	Factors that affect a population regardless of its size.	Natural events (e.g. earthquakes, storms).

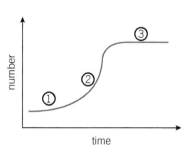

▲ **Figure 1** *A typical population growth curve. 1 = slow growth (lag phase). 2 = rapid growth. 3 = a stable state (carrying capacity); birth rate and death rate are very similar*

Synoptic link

Animal population growth curves resemble the growth curves of bacterial colonies, which you examined in Topic 22.6, Culturing microorganisms in the laboratory.

Key term

Limiting factor (in the context of populations): A factor that restricts the final size of a population.

Summary questions

1 Define **a** a limiting factor **b** a density-dependent factor. (*2 marks*)

2 Examine the demographic graph in Figure 2. Describe and explain the change in population size during stages 2 and 3. (*3 marks*)

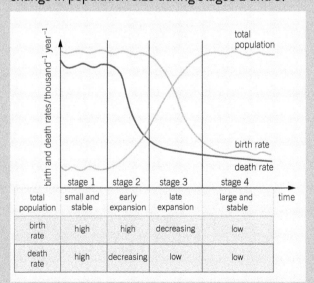

▲ **Figure 2**

3 Population pyramids illustrate the distribution of ages within populations. The two graphs in Figure 3 show age distribution in two countries with similar populations. State which country has the faster population growth. Explain your answer. (*2 marks*)

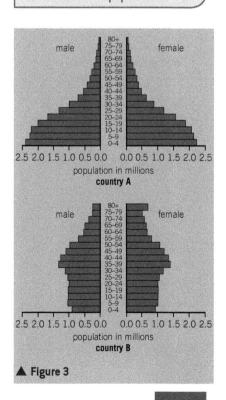

▲ **Figure 3**

24.2 Competition
24.3 Predator–prey relationships
Specification reference: 6.3.2 (b)

Members of a population compete with each other for limiting factors such as food, light, and space. Populations also interact with other species by competing for resources and through predator–prey relationships.

Intraspecific and interspecific competition

Interspecific competition = **different species** competing for resources. The competing species overlap in their food sources, behaviour, or the habitat they occupy. The better adapted species may **outcompete** the other species if resources are limited.

Intraspecific competition = members of the **same species** competing for limited resources.

Revision tip: Interspecific vs. intraspecific

It is important to specify the type of competition that is occurring. Remember that 'interspecific' means 'between species' (think of 'intervals' or 'inter-city trains' perhaps), and 'intraspecific' means 'within species'.

> ### + Go further: Allelopathy – competition between plants
>
> Some plant species enhance their ability to compete with other plants by releasing toxic or inhibitory chemicals into their surroundings, which is called **allelopathy**. Examples of this chemical warfare include the release of substances that inhibit chlorophyll production and molecules that reduce the rate of respiration. The black walnut, *Juglans nigra*, has several allelopathic tools at its disposal. It releases a chemical called juglone into the soil, which inhibits respiration in nearby plants, and also contains allelopathic chemicals in its leaves and roots.
>
> 1 Suggest how the storage of toxic chemicals in leaves can result in allelopathy.
>
> 2 Suggest why the species growing in a habitat should be assessed for their allelopathic properties before the land is selected for agriculture.

Summary questions

1 State three factors for which plant species are likely to compete. *(3 marks)*

2 Corals, limpets, and anemones are all filter feeders (they strain food particles from water) with habitats on the ocean floor. Barnacles are filter feeders that have evolved the ability to form colonies on the skin of whales and the side of ships. Explain the benefit to barnacles of this ability. *(3 marks)*

3 Analyse Figure 1 and explain why the peaks in predator numbers are lower than those of prey, and explain why the peaks are delayed compared to the peaks in prey numbers. *(3 marks)*

Analysing predator–prey relationships

All predator–prey relationships show a similar pattern: numbers in both populations fluctuate, but the peaks and troughs in the predator population are delayed in comparison to those of the prey.

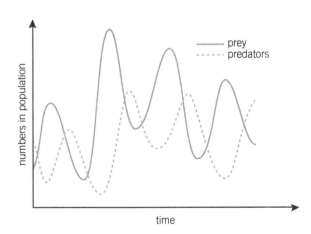

▲ **Figure 1** *A typical predator–prey graph*

24.4 Conservation and preservation
24.5 Sustainability
Specification reference: 6.3.2 (c) and (d)

Ecosystems can be protected through **conservation** or **preservation**. These two terms are often used interchangeably, but they have distinct meanings.

Conservation or preservation?

▼ **Table 1** *A comparison of conservation and preservation*

	Definition	What is done?
Conservation	The sustainable management of ecosystems to maintain biodiversity.	Active, **sustainable** management (i.e. balancing the maintenance of biodiversity with the extraction of resources). This can include **reclamation** (i.e. restoring damaged ecosystems).
Preservation	The maintenance of ecosystems in their original state (without interference).	Ecosystems are monitored, but visitors are not allowed, and interference is kept to a minimum.

The importance of conservation

Conservation is important for a range of ethical, social, and economic reasons.

Sustainability

Ecological sustainability is the exploitation of resources without compromising biodiversity or the ability to meet future requirements (i.e. ensuring that resources are renewable and will not run out).

▼ **Table 2** *Examples of sustainable practices*

	How is it done sustainably?
Timber production	**Coppicing** (for small-scale production), which involves trees being cut close to the ground. The trees remain alive and eventually produce new shoots. Coppicing can be **rotational** (i.e. the area being coppiced varies each year). Biodiversity remains high because succession is stopped. On a larger scale, trees are felled and will not regrow. To ensure sustainability, trees are **replanted** and only the largest trees are cut each year.
Fishing	Fishing **quotas** set limits on the number of fish of certain species that can be caught. Sustainability can also be improved through the use of **fish farms** (rather than catching wild fish) and promoting the consumption of species that are lower in the food chain.

Summary questions

1 Describe the difference between conservation and preservation.
(2 marks)

2 Tilapia fish are primary consumers. Salmon are tertiary consumers. Explain why the farming of tilapia is more sustainable than the farming of salmon.
(2 marks)

3 Suggest why reclamation of a habitat is difficult. *(2 marks)*

Question and model answer: The importance of conservation

Q. Describe the ethical, social, and economic benefits of conserving ecosystems.

A.

> Try to avoid vague answers such as "we should not play God" or "animals have the right to live".

Ethical – Humans have a **moral responsibility** to maintain biodiversity. (However, the ethical argument for conservation is subjective.)

Social – Conserved ecosystems provide natural beauty and are **aesthetically attractive**. They can provide amenities/**recreation**, enable **ecotourism**, and have an **educational** benefit.

Economic – Organisms can be harvested from ecosystems using a **sustainable** approach. Humans gain resources such as **food, drugs**, and timber. The maintenance of biodiversity increases the chance of discovering organisms with useful properties that have not yet been recognised.

Key terms

Conservation: Protection of an ecosystem through active management.

Preservation: Protection of an ecosystem by preventing human use or interference.

Sustainability: The use of a natural resource without damaging biodiversity and ensuring that the resource is not depleted.

24.6 – 24.9 Ecosystem management

Specification reference: 6.3.2 (e)

Ecosystems often require management to balance the maintenance of biodiversity with human economic and social needs. Here you will learn about how this potential conflict is being dealt with in a range of different ecosystems.

Ecosystem management case studies

Ecosystem	Description of the ecosystem	Conservation methods
Masai Mara, Kenya	A savannah ecosystem, containing grassland and woodland. Large populations of zebra, buffalo, elephants, leopards, and lions.	Promotion of **ecotourism** (which attempts to balance conservation with tourism, as well as benefitting local people). Prevention of rhino **poaching** (e.g. by employing a nature reserve ranger). **Legal hunting** is allowed in some cases, which culls animals considered to be in excess.
Terai region, Nepal	High temperatures and humidity in the summer months. Fertile soils with lush forest. High biodiversity. Heavy deforestation and agricultural use.	**Sustainable forest management** (e.g. harvesting quotas). Improved **irrigation** schemes (to increase the efficiency of agricultural production). Encouragement of fruit and vegetable growing in nearby regions to **relieve the pressure of intensive farming** on the Terai region.
Peat bogs	Ground with high water content and decomposing vegetation. Natural stores of CO_2. Peat can be removed for use as a fuel to improve farm soils. Contains rare species.	**Blocking ditches** that have been constructed to prevent flooding on neighbouring land. This **raises water levels** in the peat bog. Removal of trees and **prevention of afforestation** (because trees have a high water demand and reduce the water content of bogs).
The Galapagos islands	An **environmentally sensitive ecosystem** (i.e. especially vulnerable to change). Contains unique species (e.g. the Galapagos giant tortoise).	Limiting and managing tourism, with park rangers to monitor conservation projects. Strict controls over the movement of introduced animals (e.g. goats and pigs).

Summary questions

1 Describe what is meant by an environmentally sensitive ecosystem.
(1 mark)

2 Local tribes in the Masai Mara have traditionally grazed livestock using semi-nomadic farming, which involves moving area as the climate changes. Nowadays, local tribes are restricted to the edges of the nature reserve. Suggest the impact this may have on the ecosystem.
(3 marks)

3 Suggest why both conservation and preservation of peat bogs is difficult to achieve.
(3 marks)

Revision tip: Applying conservation principles

Most principles of conservation and ecosystem management apply to all the case studies outlined here. You should be prepared to apply these ideas to other ecosystems, but try to learn the specific details of the examples on this page.

1 Which of the following statements represents a density-independent factor that can influence population size? (*1 mark*)

   A Competition

   B The spread of disease

   C Climate

   D Parasitism

2 Describe how the rise in human population has affected abiotic factors in ecosystems and explain the impact these changes have had on biotic factors. (*6 marks*)

3 The graph below illustrates the effects of intraspecific competition over time. Describe what is occurring at stages 1–3. (*3 marks*)

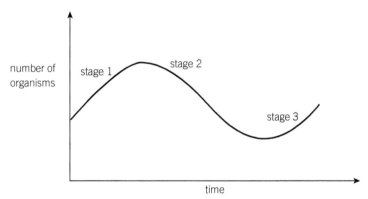

4 The graph shows a general predator–prey graph. Explain the changes to prey and predator populations at stages 1–4. (*4 marks*)

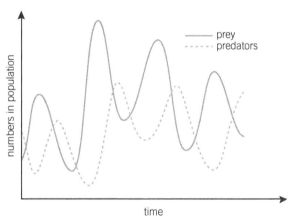

Answers to practice questions

Chapter 2

1 D *[1]*,   **2** A *[1]*,   **3** A *[1]*

4

Structure	Membrane-bound	Contains DNA	
Mitochondrion	YES	YES	*[1]*
Chloroplast	YES	YES	*[1]*
Lysosome	YES	NO	*[1]*
Microtubules	NO	NO	*[1]*

5 Fixing/fixation; *[1]*,   Differential; *[1]*,
Crystal violet; *[1]*,   Cell walls *[1]*

6 9×10^{-5} m
(1 mark awarded for $90\,\mu$m)   *[2]*

7 Both consist of microtubules;   *[1]*
Eukaryotic flagella have a 9 + 2 arrangement of
microtubules;   *[1]*
Enable cell movement in both types of cell;   *[1]*
Thinner in prokaryotes   *[1 (max 3)]*

Chapter 3

1 B *[1]*,   **2** D *[1]*,   **3** A *[1]*,   **4** B *[1]*

5 red; *[1]*,   cuvette; *[1]*,   Benedict's; *[1]*,   precipitate *[1]*

6 a Phosphate (PO_4^{3-}) ions; *[1]*,   From soil; *[1]*,
Diffusion into roots *[1]*

b Level 3 (5–6 marks): Discussion of at least
three of the following examples: phospholipids/
cell membranes, nucleic acids, ATP, NADP,
bones. Structure has been linked to function.

Level 2 (3–4 marks): Discussion of at least two
of the following examples: phospholipids/cell
membranes, nucleic acids, ATP, NADP, bones.
Both structure and function of the relevant
molecules have been discussed to some extent.

Level 1 (1–2 marks): Discussion of at least one
of the following examples: phospholipids/cell
membranes, nucleic acids, ATP, NADP, bones. The
information is supported by limited evidence.

7 a Sweating / panting;   *[1]*
Water has a high heat of vaporisation;   *[1]*
A relatively large amount of energy is required
to change water from liquid to gas;   *[1]*
The evaporation of water from the surface of an
organism has a cooling effect   *[1]*

b Dipoles shown (i.e. partial positive charge on H
atoms and partial negative charge on O atoms
within OH groups);   *[1]*
Hydrogen bond drawn between H in water and
O in ethanol (or O in ethanol and H in water) *[1]*

8 a Detergent: breaks down cell membranes;   *[1]*
Salt: breaks hydrogen bonds between water
and DNA;   *[1]*
Protease: breaks down proteins (e.g. histones);*[1]*
Alcohol: precipitates DNA   *[1]*

b i purine, pyrimidine, purine (A,C,G);   *[1]*
pyrimidine, pyrimidine, purine (T,T,A)   *[1]*

ii two *[1]*   **iii** UGC; *[1]*   AAU   *[1]*

Chapter 4

1 D *[1]*,   **2** C *[1]*,   **3** A *[1]*,   **4** C *[1]*

5 a 7   *[1]*

b Place % in the column heading rather than
in every box; *[1]* Write pH values to the same
number of decimal places   *[1]*

c Test a greater range of pH values   *[1]*

6 Complementary; *[1]*   Active site; *[1]*   Similar *[1]*

7 pH; *[1]*, Temperature; *[1]*, Enzyme concentration; *[1]*,
The overall volume of solution *[1 (max 3)]*

Chapter 5

1 C *[1]*,   **2** A *[1]*,   **3** C *[1]*,   **4** B *[1]*

5 *Cell communication*
Glycoproteins;   *[1]*
Embedded/AW in the membrane bilayer;   *[1]*
Act as receptors   *[1]*
Exchange of substances
Partially permeable;   *[1]*
Membranes surround organelles;   *[1]*
Carrier/channel proteins;   *[1]*
Only specific substances are allowed to pass
through a membrane   *[1 (max 5)]*

6 Membranes are fluid because components
(phospholipids, cholesterol, and proteins) can
move within the bilayer;   *[1]*
Membranes are compared to mosaics because
proteins are scattered throughout the bilayer, like
the tiles in a mosaic   *[1]*

7 Highest/er; *[1]*,   Lowest/er; *[1]*
Turgor; *[1]*,   Plasmolysis *[1]*

8 Water diffuses from cell B to A and C;   *[1]*
Water diffuses from cell A to C;   *[1]*
Water moves from high to low water potential   *[1]*

Chapter 6

1 B *[1]*,   **2** B *[1]*,   **3** C *[1]*

4 **Level 3 (5–6 marks):** Exemplification of stem cell therapy, including discussion of different stem cell sources (e.g. multipotent stem cells, pluripotent ESCs, induced PSCs) and detailed discussion of problems arising from research and clinical trials. There is a well-developed line of reasoning which is clear and logically structured. The information presented is relevant and substantiated.

Level 2 (3–4 marks): Exemplification of stem cell therapy (e.g. use of multipotent stem cells to treat leukaemia) and some of the problems arising from research and trials. There is a line of reasoning presented with some structure. The information presented is in the most part relevant and supported by some evidence.

Level 1 (1–2 marks): Simple comments about stem cell use with little exemplification or detail. The information is basic and communicated in an unstructured way. The information is supported by limited evidence.

5 Chromatin;   *[1]*

Nucleolus;   *[1]*

Centrioles;   *[1]*

Microtubules   *[1]*

6 A = Palisade cell (Z);   *[1]*

Explanation: greater rate of carbon dioxide diffusion;   *[1]*

B = Sperm cell (Y);   *[1]*

Explanation: the acrosome contains enzymes that enable the sperm cell to enter a secondary oocyte;   *[1]*

C = Erythrocyte (X);   *[1]*

Explanation: more space for haemoglobin to be stored;   *[1]*

D = Neutrophil (W);   *[1]*

Explanation: lysosomes contain enzymes that digest pathogens   *[1]*

Chapter 7

1 A *[1]*,   **2** C *[1]*,   **3** C *[1]*,   **4** A *[1]*

5 **a**   Radius = 0.5 µm;   *[1]*

Surface area = 3.14 µm²;   *[1]*

Volume = 0.52 µm³;   *[1]*

Ratio = 6:1   *[1]*

b   Large surface area: volume ratio;   *[1]*

Exchange can occur by diffusion alone   *[1]*

6 External; *[1]*,   Flattens; *[1]*,

Thorax/chest cavity; *[1]*,   Pressure *[1]*

Chapter 8

1 D *[1]*,   **2** A *[1]*,   **3** B *[1]*,   **4** D *[1]*,   **5** C *[1]*

6 Plasma; *[1]*,   Hydrostatic; *[1]*,

Potential; *[1]*,   Venous *[1]*

7 **a**   X = endothelium;   *[1]*

Y = (smooth) muscle   *[1]*

b   Flexibility; *[1]*   Enabling stretch and recoil   *[1]*

Chapter 9

1 C *[1]*,   **2** D *[1]*,   **3** B *[1]*,   **4** C *[1]*

5 Root pressure drops when oxygen levels decrease;   *[1]*

Temperature and root pressure are correlated;   *[1]*

Respiratory poisons (e.g. cyanide) reduce/stop root pressure;   *[1]*

Xylem sap   *[1 (max 3)]*

6 W = cortex *[1]*,   X = phloem *[1]*,

Y = xylem *[1]*,   Z = parenchyma *[1]*

7 Mesophyll; *[1]*,   Xylem vessels; *[1]*,

Cohesion; *[1]*,   Capillary *[1]*

Chapter 10

1 C *[1]*,   **2** D *[1]*,   **3** B *[1]*

4 Mutation; *[1]*,   Alleles; *[1]*,

Change in selection pressure; *[1]*,   Adapted *[1]*

5 **a**   Normal   *[1]*

b   **i**   Continuous   *[1]*

ii   Polygenic/influenced by many genes;   *[1]*

Environmental influence   *[1]*

c   Mean, mode, and median are the same;   *[1]*

164 cm   *[1]*

6 *Level 3 (5–6 marks)*: Detailed and accurate descriptions of the types of evidence used in classification, with examples to illustrate each type of evidence. There is a well-developed line of reasoning which is clear and logically structured. The information presented is relevant and substantiated.

Level 2 (3–4 marks): Includes some accurate descriptions of the types of evidence used in classification, with examples to illustrate most types of evidence. There is a line of reasoning presented with some structure. The information presented is in the most part relevant and supported by some evidence

Level 1 (1–2 marks): Simple comments about the evidence used in classification, with few correct examples. The information is basic and communicated in an unstructured way. The information is supported by limited evidence.

Relevant scientific points include:

Biochemical/molecular evidence (including DNA and amino acids);

Some candidates may include details of the techniques used to analyse these molecules;

Examples include: cytochrome c differences between species; comparisons of hominid DNA in phylogeny.

Anatomical evidence;

Candidates may discuss limitations of anatomical evidence, including convergent evolution;

Immunological evidence;

Behavioural evidence; examples may include similarities in primate behaviour;

Embryological evidence; examples include similarities in vertebrate embryos

Chapter 11

1 a D; *[1]*

(11/33) × 100 = 33.3% of genes are polymorphic; *[1]*

Population D has the highest proportion of polymorphic genes *[1]*

b (Reliability is doubtful because) only a small proportion of the genome has been sampled; *[1]*

Different numbers of genes from each population have been analysed *[1]*

2 a Habitat, destruction/degradation/change/loss; *[1]*

Climate change; *[1]*

(Introduction of) invasive species; *[1]*

Pollution *[1 (max 2)]*

b i 4 965 000 *[1]*,   **ii** 4 896 000 *[1]*

c **Level 3 (5–6 marks)**: Detailed description of potential sampling methodology and suggestions for approaches to analysis of the data. There is a well-developed line of reasoning which is clear and logically structured. The information presented is relevant and substantiated.

Level 2 (3–4 marks): Generally accurate description of potential sampling methodology and suggestions for approaches to analysis of the data. There is a line of reasoning presented with some structure. The information presented is in the most part relevant and supported by some evidence.

Level 1 (1–2 marks): Simple comments about either sampling methodology or analysis. The information is basic and communicated in an unstructured way. The information is supported by limited evidence.

Relevant scientific points include:

Sampling

Larger sampling area / number of samples improves validity;

Avoiding bias when sampling / random sampling;

Description of stratified sampling;

Suitable (repeatable) method for counting a mammalian species;

Regular sampling over time (several years) to monitor trends

Analysis

Use of statistical tests to analyse trends over time;

Analysis of genetic diversity within the species (e.g. heterozygosity or proportion of polymorphic alleles);

Estimation of total population size from samples;

Calculation of rate of decline (of population);

Calculation of range / distribution of species

3 C *[1]*,   **4** A *[1]*,   **5** B *[1]*

Chapter 12

1 B *[1]*,   **2** B *[1]*,   **3** C *[1]*

4 a 118 *[1]*,   **b** 20 *[1]*

5 Danger of a live vaccine reactivating; *[1]*

HIV infects lymphocytes and evades immune detection; *[1]*

High mutation rate *[1 (max 2)]*

6 a Leave space between plants; *[1]*

Control / kill insect vectors; *[1]*

Rotate crops (to limit contamination via soil); *[1]*

Follow hygiene practices *[1 (max 3)]*

b Regular hand washing; *[1]*

Efficient waste disposal; *[1]*

Washing of bedding (to reduce fomite transmission) *[1 (max 2)]*

7 Receptors; *[1]*,   Nucleus; *[1]*,   Callose; *[1]*,   Phloem; *[1]*,   Lignin *[1]*

Chapter 13

1 D *[1]*,   2 A *[1]*,   3 D *[1]*,   4 D *[1]*,
5 C *[1]*

6 **Level 3 [5–6 marks]:** Detailed comparison of the function and physiology of the two branches of the nervous system. There is a well-developed line of reasoning which is clear and logically structured. The information presented is relevant and substantiated.

Level 2 [3–4 marks]: Includes some correct comparisons of the two branches of the nervous system. There is a line of reasoning presented with some structure. The information presented is in the most part relevant and supported by some evidence.

Level 1 [1–2 marks]: Simple comments about the two branches. The information is basic and communicated in an unstructured way. The information is supported by limited evidence.

Relevant scientific points include:

Differences in neurotransmitters, ganglion position, general function (sympathetic = 'flight, fright, fight, or excite'; parasympathetic = 'rest and digest'), and examples of contrasting specific functions (e.g. sympathetic speeds up heart rate whereas parasympathetic slows heart rate).

Chapter 14

1 B *[1]*,   2 C *[1]*,   3 C *[1]*

4 **Level 3 [5–6 marks]:** Detailed description of both hormonal and nervous control of heart rate, with use of appropriate terminology. There is a well-developed line of reasoning which is clear and logically structured. The information presented is relevant and substantiated.

Level 2 [3–4 marks]: Accurate description of aspects of hormonal and nervous control. There is a line of reasoning presented with some structure. The information presented is in the most part relevant and supported by some evidence.

Level 1 [1–2 marks]: Simple comments about heart rate control. The information is basic and communicated in an unstructured way. The information is supported by limited evidence.

Relevant scientific points include:

adrenaline increases heart rate / stroke volume / cardiac output;

cardiovascular centre in medulla oblongata;

idea of nervous connection to SAN / sino-atrial node;

(which) controls frequency of waves of excitation / depolarisation;

vagus / parasympathetic , nerve decreases heart rate;

accelerator / sympathetic , nerve increases heart rate;

high blood pressure detected by stretch receptors / baroreceptors;

low blood pH / increased levels of blood CO_2 detected by chemoreceptors;

(receptors) in , aorta / carotid sinus / carotid arteries.

5 (Adrenaline is a) first messenger;

Binds to receptors on cell surface membranes;

cAMP is a second messenger;

(triggers intracellular) cascade effect;

Glycogenolysis;

Glucose is produced and diffuses out of cells   *[max 5]*

6 *[3]*

Trait	Type 1	Type 2
Treatment	Insulin injections	Dietary changes
Extent of genetic influence	Large / AW	Some / AW
Is insulin produced?	No / very little	Usually (but often less than normal)

Chapter 15

1 A *[1]*,   2 B *[1]*,   3 D *[1]*
4 A *[1]*,   5 A *[1]*,   6 C *[1]*

Chapter 16

1 D *[1]*,   2 A *[1]*

3 Some amylase is produced without gibberellin exposure;

Positive correlation;

Between gibberellin exposure and amylase production;

Amylase production plateaus after 13 days of gibberellin exposure;

Gibberellin triggers amylase production;

By switching on the gene coding for amylase

[max 5]

4 Auxin / one plant hormone can produce several effects depending on concentration;

Opposite effects can be produced by altering concentration;

Precise measurements of concentration are needed for tissue culture;

Concentrations will vary depending on the species;

Concentrations will vary depending on the stage of the culture   *[max 2]*

Chapter 17

1 D *[1]*,   2 D *[1]*,   3 A *[1]*

4 **Level 3 [5–6 marks]:** Detailed and accurate descriptions of the methods that can be used to improve primary productivity, with examples to illustrate each method. There is a well-developed line of reasoning which is clear and logically structured. The information presented is relevant and substantiated.

Level 2 [3–4 marks]: Includes some accurate descriptions of the methods used to manipulate productivity. There is a line of reasoning presented with some structure. The information presented is in the most part relevant and supported by some evidence.

Level 1 [1–2 marks]: Simple comments about the methods used on farms to improve productivity, with few correct examples. The information is basic and communicated in an unstructured way. The information is supported by limited evidence.

Relevant scientific points include:

Consideration of light banks, sowing timing, sowing density, greenhouse temperature regulation, irrigation, crop rotation, fertilisers, herbicides, pesticides.

5 Initially the increase in the rate of photosynthesis is proportional to the increase in light intensity (i.e. linear relationship);

At this point, light intensity is a limiting factor;

The graph plateaus at a high light intensity;

At this point, CO_2 concentration is a limiting factor
[max 3]

Chapter 18

1 C *[1]*,   2 B *[1]*,   3 D *[1]*,   4 B *[1]*

5 **Level 3 [5–6 marks]:** Detailed and accurate descriptions of the use of respirometers and the design of an experiment to measure RQ. There is a well-developed line of reasoning which is clear and logically structured. The information presented is relevant and substantiated.

Level 2 [3–4 marks]: Includes some accurate descriptions of the use of respirometers and valid experimental design. There is a line of reasoning presented with some structure. The information presented is in the most part relevant and supported by some evidence.

Level 1 [1–2 marks]: Simple comments about the use of respirometers and experimental design. The information is basic and communicated in an unstructured way. The information is supported by limited evidence.

Relevant scientific points include:

Details of the respirometer, including the role of soda lime, potassium hydroxide, the double chamber design, index liquid.

Details of a valid experimental design, including the use of a control respirometer with glass beads, control variables, measurement of the meniscus position, repeat readings.

An understanding of RQ calculation.

6 In respiration, small amounts of energy are released;

In a series of chemical reactions;

This enables the transfer of chemical energy to bonds in ATP;

In combustion, energy is released rapidly/in one go
[max 3]

Chapter 19

1 A *[1]*

2 a Lac operon *[1]*

   b Lactose binds to the repressor protein;

   Alters the shape of the repressor protein;

   Prevents the repressor protein binding to the operator   *[max 2]*

3 Programmed;

Enzymes;

Blebs;

Phagocytes   *[max 4]*

4 Post-transcriptional/pre-translational (control);

Splicing;

Variation in which introns are retained and removed can produce different polypeptides;

Exons can be rearranged (to produce different polypeptides);   *[max 3]*

5 A = Substitution   B = Silent

C = Missense   *[max 3]*

Chapter 20

1 B *[1]*,   2 C *[1]*,   3 B *[1]*,   4 D *[1]*

5 C *[1]*

6 **Level 3 [5–6 marks]:** Accurate explanation of the founder effect with at least two detailed examples from populations. There is a well-developed line of reasoning, which is clear and logically structured. The information presented is relevant and substantiated.

Level 2 [3–4 marks]: Includes a description of the founder effect and one example from populations. There is a line of reasoning presented with some structure. The information presented is in the most part relevant and supported by some evidence.

Level 1 [1–2 marks]: Simple comments about the founder effect. The information is basic and communicated in an unstructured way. The information is supported by limited evidence.

Relevant scientific points include:

Genetic bottlenecks, reduced genetic diversity, examples such as Ellis–van Creveld syndrome, blood group distribution.

Chapter 21

1 B *[1]*,   2 C *[1]*,   3 D *[1]*

4 **A** DNA polymerase / Taq polymerase

   B Restriction endonucleases

   C (DNA) ligase

   D Plasmid(s)

   E Reverse transcriptase   *[max 5 marks]*

5 Genetic engineering = C

  Somatic cell therapy = D

  DNA profiling = B

  Animal reproductive cloning = A   *[max 4 marks]*

Chapter 22

1 Propagation; *[1]*   Meristem; *[1]*

  Suckers; *[1]*   Genetically; *[1]*

2 D, B, C, A   *[4]*

3 Finite / limited glucose supply;   *[1]*

  Glucose used up during the log phase;   *[1]*

  Penicillin produced in stationary phase;   *[1]*

  Secondary metabolites produced in times of stress / not produced as part of normal metabolism   *[1]*

4 (From left to right)

  Adsorption (to inorganic material / cellulose / silica); matrix entrapment; covalent or ionic bonding to inorganic carrier; membrane entrapment / encapsulation   *[max 4]*

Chapter 23

1 D *[1]*

2 **a** Oak (leaves) and nettle (leaves)   *[1]*

  b Sparrowhawk, badger, owl, blue tit *[1]*

  c Blue tit population would reduce/disappear;

     Aphid population would rise;

     Nettle population would be reduced;

     Sparrowhawk population would be reduced;

     Owl population would be reduced;

     Blackbird population may reduce (as sparrowhawks have fewer blue tits on which to predate);

     Shrew population may reduce (as owls have fewer blue tits on which to predate);

     Badger population may decline   *[max 4]*

d Shrew to badger / shrew to owl / blackbird to sparrowhawk / blue tit to sparrowhawk;

   More energy released as heat / more (named) indigestible parts   *[2]*

3 **a** 4% *[1]*   **b** 20% *[1]*   **c** 15% *[1]*

4 Nitrification = A, B, D   *[1]*

  Denitrification = C, G   *[1]*

  Nitrogen fixation = E, F   *[1]*

Chapter 24

1 C *[1]*

2 **Level 3 [5–6 marks]:** Detailed descriptions of abiotic changes, with clear links to the subsequent effects on biodiversity and the use of examples. There is a well-developed line of reasoning which is clear and logically structured. The information presented is relevant and substantiated.

  Level 2 [3–4 marks]: Includes some accurate descriptions of abiotic changes, with some links to the effects on biodiversity. There is a line of reasoning presented with some structure. The information presented is in the most part relevant and supported by some evidence.

  Level 1 [1–2 marks]: Simple comments about abiotic changes or biotic changes, with few correct examples. The information is basic and communicated in an unstructured way. The information is supported by limited evidence.

  Relevant scientific points include:

  Effects on climate (e.g. climate change), soils (e.g. salinity changes), and water quality (e.g. eutrophication).

  Impact of climate change on vulnerable species (e.g. frog species), deforestation causing extinction/ reduced biodiversity, the impact of pesticides on some species.

3 Stage 1: resources are plentiful and population size increases;   *[1]*

  Stage 2: Resources become limited and population begins to decrease;   *[1]*

  Stage 3: Competition is reduced and the population rises again;   *[1]*

4 Stage 1: Prey population rises, which provides more food for predators;   *[1]*

  Stage 2: Prey population declines because of increased predation;   *[1]*

  Stage 3: Predator population decreases because of increased intraspecific competition for reduced prey numbers;   *[1]*

  Stage 4: Prey population rises again because predator numbers have decreased.   *[1]*

Answers to summary questions

2.1

1 To enable light to pass through from below   *[1]*

2 Objective magnifies the image;   *[1]*

   Eyepiece increases the magnification further   *[1]*

3 Differential staining (described);   *[1]*

   Xylem will stain yellow;   *[1]*

   Non-lignified cell walls will stain blue   *[1]*

2.2

1 *Magnification*: size of image divided by actual size of object;   *[1]*

   Resolution: the shortest distance between two objects that can still be distinguished as separate structures   *[1]*

2 1.5;   *[1]*

   μm   *[1]*

   (or 0.0015 mm)

3 0.0112 m (or 1.12 cm);   *[1]*

   1.12×10^{-2} m (in standard form);   *[1]*

   1.1×10^{-2} m (to 2 significant figures)

   [1 (the correct answer in standard form receives all three marks without working)]

2.3

1 Electron microscopes have greater resolution;   *[1]*

   Because the wavelengths of electrons are shorter than the wavelengths of light   *[1]*

2 0.30 m (or 300 000 μm);   *[1]*

   3.0×10^{-1} m

   [1 (the correct answer in standard form receives both marks without working)]

3 *Electron microscopes*:

   Greater resolution;   *[1]*

   Higher magnification;   *[1]*

   Specimen must be dead (whereas CLSMs can use living tissue);   *[1]*

   Both:

   3D images (but not with TEMs);   *[1]*

   Expensive/artefacts can be introduced in preparation   *[1]*

   (accept reverse arguments)

2.4

1 Both contain microfilaments (in a 9 + 2 arrangement);   *[1]*

   Flagella are longer than cilia   *[1]*

2 Synthesis at rough ER/ribosomes;   *[1]*

   Transported to Golgi in vesicle;   *[1]*

   Fusion of vesicle with Golgi membrane;   *[1]*

   Modified within Golgi;   *[1]*

   Packaged into secretory vesicle;   *[1]*

   Fusion of vesicle within plasma cell membrane   *[1 (max 5)]*

3 Large networks help to control movements (e.g. of organelles) within cells;   *[1]*

   Rapid assembly and disassembly can produce movement of whole cells (i.e. controls cell migration);   *[1]*

   (enables rapid) cytokinesis/cell division   *[1]*

2.5

1 Maintaining turgor   *[1]*

2 Other groups of organisms contain cell walls and vacuoles (other than some algal species, only plants contain chloroplasts);   *[1]*

   The cell walls and vacuoles in other groups, however, exhibit structural differences to plant cells (e.g. bacterial cell walls contain peptidoglycan);   *[1]*

   Not all plant cells contain chloroplasts   *[1 (max 2)]*

3 Chloroplasts may have evolved from bacteria;   *[1]*

   Small bacteria may have entered larger cells (i.e. symbiosis);   *[1]*

   Ribosomes in chloroplasts have evolved a different structure to prokaryotic ribosomes, suggesting protein synthesis works differently in chloroplasts and bacteria   *[1]*

2.6

Go further

1 The artefacts were probably caused by damage to the plasma membrane during the preparation of the cells for use in electron microscopes

2 Aerobic respiration does occur on the plasma membrane of bacteria, but without infoldings

Summary questions

1 Prokaryotic walls are made of peptidoglycan/murein;   *[1]*

   Eukaryotic walls are made of chitin/cellulose   *[1]*

2 Increases genetic variation;   *[1]*

   Bacteria reproduce asexually, which limits variation   *[1]*

3 *Eukaryotic*

   Linear chromosomes;   *[1]*

   Histone proteins;   *[1]*

Within a nucleus; [1]

Also found in mitochondria/chloroplasts; [1]

Prokaryotic

Naked/no nucleus; [1]

Circular; [1]

Also found in plasmids [1 (max 4)]

3.1

1 Carbon; [1]

Oxygen; [1]

Hydrogen [1]

2 *Compound*: more than one element bonded together; [1]

Molecule: more than one atom; [1]

covalently bonded [1]

3 a Cl^-, HCO_3^-, NH_4^+, Ca^{2+}, OH^- [1]

   b CH_3COO^- [1]

   c HCO_3^-, NH_4^-, CH_3COO^-, OH^- [1]

3.2

1 Oxygen attracts electrons more than hydrogen; [1]

Oxygen has a negative dipole/partial charge; [1]

Hydrogen has a positive dipole/partial charge; [1]

2 Hydrogen bonds; [1]

Water molecules have cohesion and flow together; [1]

Water is a good solvent/dissolves and transports important biological molecules and ions [1]

3 The positive charges are attracted to the δ^- O atom in water; [1]

The negative charges are attracted to the δ^+ H atom in water [1]

3.3

1 Condensation reaction; [1]

Between glucose and fructose; [1]

Glycosidic bond formed [1]

2 Both contain C, O, and H only; [1]

Both have ring structures; [1]

Glucose has 6 C atoms/ribose has 5 C atoms [1]

3 Cellulose is made from β-glucose monomers; [1]

Glycogen is made from α-glucose monomers; [1]

Cellulose has hydrogen bonds/cross-links between chains; [1]

Cross-links increase strength, which is required in cellulose's support role in cell walls; [1]

Glycogen has 1,6 glycosidic bonds, which enables branching; [1]

Branching makes glycogen compact; [1]

Branching increases the number of points at which glucose monomers can be hydrolysed; [1]

Both polymers are insoluble [1 (max 6)]

3.4

1 Colorimetry (following the Benedict's test); [1]

Biosensor measurements; [1]

Reagent strips [1]

2 a *absorption*: glucose concentration on the x-axis and absorption on the y-axis; [1]

Absorption decreases as glucose concentration increases; [1]

Explanation: more Benedict's reagent reacts as glucose concentration increases; [1]

Less light is absorbed (and more passes through) when less Benedict's reagent (which absorbs red light) remains; [1]

   b *transmission*: glucose concentration on the x-axis and transmission on the y-axis; [1]

Transmission increases as glucose concentration increases [1]

3 Benedict's reagent is blue and therefore absorbs red light; [1]

Measurements of transmission will be more precise if only red light is allowed to reach the solution [1]

3.5

1 White/creamy; [1]

emulsion [1]

2 Hydrophilic phosphate; [1]

Faces/interacts with water (in tissue fluid and cytosol); [1]

Hydrophobic fatty acids; [1]

Form a partially permeable barrier [1]

3 *Similarities*: condensation reactions; [1]

Water produced; [1]

Enzymes catalyse the reactions; [1]

Differences: polypeptides are polymers, but triglycerides are not; [1]

Peptide bonds formed in polypeptides, but ester bonds formed in triglycerides; [1]

Triglycerides always consist of glycerol and three fatty acids, whereas polypeptides vary in the number of amino acids they contain [1 (max 4)]

3.6

Go further

1 They are hydrophobic

2 More intermolecular bonds form between glutenins and gliadins

3 The salt ions interfere/block charges on proteins that might otherwise cause repulsion between polypeptides

Summary questions

1 Central C atom; *[1]*

Four groups bonded to the central carbon: COOH, NH_2, H, and CH_3 *[1]*

2 a 0.83 *[1]*

b 3.55 cm *[1]*

3 Prepare a range of protein solutions of known concentration; *[1]*

Add a specific volume of biuret solution to each protein solution; *[1]*

Use colorimetry to plot a calibration curve (protein concentration against absorption; *[1]*

Test the unknown sample and calculate protein concentration from the calibration curve *[1]*

3.7

1 Golgi apparatus *[1]*

2 Globular proteins often need to be transported (e.g. in blood) and therefore need to be water-soluble; *[1]*

Fibrous proteins tend to have structural roles and should not dissolve in water. *[1]*

3 Keratin's flexibility increases when the number of disulfide bonds decreases; *[1]*

Hair is more flexible than nails *[1]*

3.8

1

Monomer	Polymer	Bond formed in condensation reaction
Monosaccharide	Polysaccharide	Glycosidic
Amino acid	Polypeptide	Peptide
Nucleotide	Polynucleotide	Phosphodiester

[1 mark per correct row]

2 Complementary base pairing; *[1]*

A pairs with T, C pairs with G *[1]*

3 Molecular structures are not complementary; *[1]*

Hydrogen bonds cannot form *[1]*

3.9

1 CGG; *[1]*

TTT; *[1]*

AGA *[1]*

2 Helicase; *[1]*

For unwinding and unzipping DNA double helix; *[1]*

DNA polymerase; *[1]*

For joining sugar-phosphate backbone/nucleotides within a new strand *[1]*

3 Degenerate code; *[1]*

The same amino acid might still be coded for even if one base changes within a codon (i.e. a substitution/point mutation) *[1]*

3.10

1 Antisense *[1]*

2 DNA can be replicated; *[1]*

Increases stability/reduces damage; *[1]*

Reduces the possibility of unwanted bonding within a single strand *[1 (max 2)]*

3

	DNA replication	Transcription	Translation
Hydrogen bonds between complementary DNA base pairs are broken	Yes	Yes	No
Free DNA nucleotides are activated	Yes	Yes	No
To which base does adenine bond?	T	U	U
Phosphodiester bonds are formed	Yes	Yes	No
Peptide bonds are formed	No	No	Yes
Location	Nucleus	Nucleus	Ribosomes (rough ER)
Product	DNA	mRNA	Polypeptide

[1 mark per correct row]

3.11

1 Both contain ribose; *[1]*

Both contain phosphates and a base; *[1]*

ATP contains three phosphates/RNA nucleotides contain one phosphate; *[1]*

The base in ATP is always adenine/RNA nucleotides can contain A, U, C, or G *[1]*

2 ATP is unstable; *[1]*

Because of phosphate bonds that are easily broken *[1]*

3 ATP is found in all living cells; *[1]*

Energy is 'carried' in chemical bonds *[1]*

4.1

1   (all enzymes are) proteins/polymers of amino acids *[1]*; (all enzymes have a) globular structure *[1]*; folded *[1]*; contain active sites *[1] (max 3)*

2   **a**   anabolic because a larger molecule (glycogen) is formed from smaller molecules (glucose) *[1]*;

   b   catabolic because smaller molecules (maltose) are formed from a larger molecule (amylose) *[1]*;

   c   catabolic because a smaller molecule (glycerol) is formed from a larger molecule (a triglyceride) *[1]*

3   alternative reaction pathway/AW *[1]*; temporary bonds between substrate and active site *[1]*; substrate bonds are strained/stretched *[1]*; less energy required to break the bonds in the substrate *[1]*

4.2

1   Denaturation *[1]*; secondary/tertiary structure of the enzyme is changed *[1]*; active site is altered *[1]*; active site is no longer a complementary shape to the substrate *[1]*; high temperatures cause weak bonds in the enzyme to vibrate and break *[1]*; pH changes result in hydrogen and ionic bonds breaking *[1] (max 5)*

2   The pH in the human body will be lower than the pH in the lakes *[1]*; enzymes in the microorganisms will not function at an optimal rate/may be denatured *[1]*

3   Psychrophile enzyme structures are more flexible *[1]*; different bonds in the tertiary structure *[1]*; fewer hydrogen bonds/more disulfide bonds *[1] (max 2)*

4.3

1   Competitive inhibitors bind to active sites/non-competitive inhibitors bind to allosteric sites *[1]*; non-competitive inhibitors alter the tertiary structure of enzymes *[1]*

2   Competitive inhibition *[1]*; similar shape/structure to the substrate/PABA *[1]*; complementary shape to the enzyme's active site *[1]*; competes with PABA for the active site *[1]*; fewer enzyme-substrate complexes formed *[1] (max 4)*

3   Negative feedback occurs when a change in a variable leads to the reversal of the change *[1]*; (in end-product inhibition) as the concentration of product increases, the reaction that produces the product is inhibited *[1]*;

4.4

1   Prosthetic groups bind permanently/tightly (whereas coenzymes bind temporarily/loosely) *[1]*;

all coenzymes are organic (whereas prosthetic groups can be organic or inorganic) *[1]*

2   Otherwise digestive enzymes would cause damage to cells *[1]*; the enzymes are activated in the digestive tract *[1]*

3   Translated polypeptide *[1]* moves in vesicle *[1]* to Golgi apparatus *[1]* where prosthetic group is added *[1]*

5.1

Go further

1   Cholesterol increases the range of temperatures over which a membrane can maintain an appropriate fluidity (i.e. a fluidity that enables it to function). Cholesterol resists changes to fluidity, thereby stabilising a membrane

2   Cholesterol will maintain fluidity (and therefore permeability) at cold temperatures and enable diffusion of important particles (e.g. O_2) into the bacterial cells

Summary questions

1   Compartmentalisation of organelles *[1]*; anchors/supports cytoskeleton *[1]*; a site for chemical reactions (e.g. thylakoid membranes in chloroplasts) *[1]*; vesicle formation *[1] (max 3)*

2   **a**   carrier proteins *[1]* because they enable facilitated diffusion of large molecules *[1]*

   b   glycoproteins *[1]* because the carbohydrate attached to the protein acts as a receptor *[1]*

3   Proteins could move within the membrane *[1]* and end up in the wrong positions within the membrane *[1]*

5.2

1   Increases in temperature raise membrane fluidity and therefore permeability *[1]* by increasing the kinetic energy of phospholipids *[1]*, which widens the spaces between the phospholipid molecules *[1]*

2   Proteins in the membrane are denatured *[1]*, preventing cell signalling/transmembrane transport *[1]*

3   Ethanol is lipid soluble *[1]* and can insert itself between phospholipids *[1]*, increasing membrane permeability *[1]*; extra particles can leave and enter cells *[1]*, thereby altering particle concentrations and cell content *[1]*

5.3

1   **a**   Microvilli increase surface area *[1]*, which increases diffusion rate *[1]*;

   b   For facilitated diffusion of glucose *[1]*, which is large and polar and therefore cannot diffuse through the phospholipid bilayer *[1]*

2 a Diffusion though the bilayer *[1]* because CO_2 is small and non-polar *[1]*;

b Facilitated diffusion *[1]* because potassium ions are charged and therefore not lipid soluble *[1]*

3 Surface area *[1]* × concentration gradient *[1]*/ membrane thickness *[1]*

5.4

1 Molecules being transported out of cells *[1]*, which are large/polar *[1]*; often proteins *[1]*; examples: hormones/enzymes/neurotransmitters *[1]*

2 One mark per correct row

	Active transport	Facilitated diffusion
Uses carrier proteins	Yes	Yes
Particles move down concentration gradients	No	Yes
Particles move against concentration gradients	Yes	No
Requires ATP	Yes	No
At least two binding sites must be present on carrier proteins	Yes (one for the particle, one for ATP)	No

3 Larger vesicles for phagocytosis *[1]*; phagocytosis transports large solid material into cells *[1]*; pinocytosis transports small, dissolved particles and small amounts of water into cells *[1]*

5.5

1 a False *[1]*

b True *[1]*

c False *[1]*

2 a increase in cell size (possible lysis/bursting, but this is unlikely because the difference in water potential is small) *[1]*

b No change *[1]*

c The cell shrinks/experiences crenation *[1]*

3 Water potential in the soil water is reduced *[1]*; water leaves plant (root hair) cells via osmosis *[1]*; plant cells plasmolyse *[1]* and lose turgor *[1]*

6.1

Go further

1 9.6 hours

2 Possible health problems include DNA damage, cancer, or malnourishment, such as a shortage of certain nutrients

Summary questions

1 a Organelles *[1]* (named) biochemicals *[1]*

b DNA *[1]*

2 The timing of each step in the cell cycle is crucial *[1]*; checkpoints regulate the sequence of events in the cycle *[1]*; the cycle will be halted if errors are detected *[1]*

3 a Although cells are not dividing in interphase, they are preparing to divide (by synthesising molecules and organelles) *[1]*

b Cells in G_0/outside the cell cycle/some differentiated cells *[1]* although these cells are not 'resting' (i.e. they are metabolically active despite not dividing) *[1]*

6.2

1 a 92 *[1]*

b 46 *[1]*

2 Centrioles are not essential for mitosis *[1]*; spindle fibres are produced through a different method in plants *[1]*

3 Cancer *[1]*; the normal control of cell division has broken down *[1]*

6.3

1 a Mitosis produces diploid cells from diploid cells/ meiosis II produces haploid cells from haploid cells *[1]*; genetically different daughter cells produced by meiosis II *[1]*

b meiosis I: homologous chromosomes separated during anaphase *[1]*; meiosis II: sister chromatids separated during anaphase *[1]*

2 524 288 *[1]*

3 Crossing over *[1]* (and) independent assortment of chromosomes/chromatids *[1]* create new combinations of alleles *[1]*; (independent assortment of chromatids introduces variation because) crossing over creates genetic differences between sister chromatids *[1]*

6.4

1 Mitochondria for ATP/energy *[1]*; acrosome contains digestive enzymes for entry into egg *[1]*; tail/flagellum for movement *[1]*; protein fibres strengthen tail/flagellum *[1]*

2 Thin/flattened cells *[1]*; present in capillaries (next to alveoli) *[1]* and alveoli *[1]*; reduce diffusion pathway/distance *[1]*

3 Thick cell wall *[1]*; lignin deposits for strength *[1]*; cells are packed closely *[1]*

6.5

1 Both can differentiate into any cell type *[1]*; only totipotent cells have the potential to develop into a whole organism *[1]*

2 They are obtained from adult cells *[1]*; overcomes ethical objections to embryonic stem cell use *[1]*; because iPSCs can be produced from cells of the patient, tissue rejection is overcome *[1]*

3 Mature plants have totipotent cells *[1]* known as meristem cells *[1]*; these cells are able to differentiate into a new plant *[1]*

7.1

1 Relatively high metabolic rate *[1]* requires greater oxygen supply (and carbon dioxide removal) *[1]*; low surface area to volume ratio *[1]* means the diffusion pathway is too long for diffusion alone to be effective *[1]*

2 Good blood supply (i.e. dense capillary networks) *[1]* removes oxygen (and supply carbon dioxide) at a high rate *[1]*; good ventilation *[1]* supplies oxygen (and removes carbon dioxide) at a high rate *[1]*

3 *Actinophrys* (radius = $22.5\,\mu m$): SA = $0.00636\,\mu m^2$ (6.36×10^{-9} m) *[1]*; V = $4.77 \times 10^{-14}\,m$ *[1]*; SA: V = 133,333:1 *[1]*

Actinosphaerium (radius = $200\,\mu m$): SA = $0.5024\,\mu m^2$ (5.02×10^{-7} m) *[1]*; V = $3.35 \times 10^{-11}\,m$ *[1]*; SA: V = 14,997:1 *[1]*

7.2/7.3

1 Short diffusion pathway due to flattened (squamous) epithelia in alveoli *[1]* and close proximity of alveoli and capillaries *[1]*; large surface area to volume ratio *[1]*; steep concentration gradient of oxygen and carbon dioxide *[1]* due to good blood supply (dense capillary network) *[1] (max 3)*

2 12 breaths min⁻¹ *[2]* (one mark for correct answer not rounded to two significant figures, e.g. 11.923)

3 **a** Gases in the alveoli and alveolar capillaries/ pulmonary vein are at equilibrium *[1]* because gas exchange occurs until the concentrations on either side of the alveolar walls are equal *[1]*

  b The pulmonary artery carries blood to the lungs from tissues (via the heart) *[1]*; oxygen concentration is lower (and carbon dioxide concentration is higher) than in alveolar air because respiration has happened in tissues *[1]*

7.4

1 Exoskeleton limits diffusion across body surfaces *[1]* and some insects have relatively high energy requirements (and therefore high oxygen demand) *[1]*

2 **Level 3 *[5–6 marks]*:** Detailed comparisons, with at least three examples of adaptations in each system, explained correctly and linked to function. There is a well-developed line of reasoning which is clear and logically structured. The information presented is relevant and substantiated.

Level 2 *[3–4 marks]*: Includes at least two examples of adaptation per system, explained correctly. There is a line of reasoning presented with some structure. The information presented is in the most part relevant and supported by some evidence

Level 1 *[1–2 marks]*: Simple comments about at least one type of adaptation per system. The information is basic and communicated in an unstructured way. The information is supported by limited evidence.

Relevant scientific points include:

Alveoli: large surface area increases the rate of gas exchange; thin epithelia reduce the diffusion distance; good blood supply maintains concentration gradients; surfactants prevent the collapse of alveoli.

Gills: Large surface area (due to may plated lamellae being present on many filaments) increases the rate of diffusion; thin cell layer in lamellae reduce the diffusion distance; good blood supply in lamellae maintains concentration gradient; countercurrent flow maintains concentration gradient.

3 Countercurrent flow *[1]*; blood in capillaries and water flow in opposite directions *[1]*; oxygen concentration is higher in the water than the capillaries along the length of the gills *[1]*

8.1

1 Diffusion *[1]* because of flatworms' high surface area to volume ratio *[1]*

2 Blood pressure is maintained *[1]*; oxygenated and deoxygenated blood does not mix *[1]*; lower volumes of transport fluid (blood) required *[1]*; blood supply to different tissues can be varied depending on demand *[1]*; delivery of oxygen and nutrients is more efficient *[1] (max 3)*

3 Countercurrent gaseous exchange (see Topic 7.4) in gills *[1]* improves the efficiency of oxygen delivery to respiring tissues *[1]*; body weight is supported in the water *[1]* and fish do not need to maintain their body temperature *[1]*; therefore metabolic demands are relatively low *[1] (max 3)*

8.2

1 Thin walls *[1]*; gaps between cells in the wall *[1]*; high permeability for diffusion across the wall *[1]*

2 Arteries are under higher pressure *[1]*; elastic fibres stretch when pressure is high and recoil as pressure falls *[1]*

3 **a** Veins: 1×10^{-2} m *[1]* arteries: 5×10^{-3} m *[1]*

  b Veins have wider lumens to reduce resistance to blood flow at lower pressures *[1]*

8.3

1 Tissue fluid has no large plasma proteins *[1]* and red blood cells *[1]*; tissue fluid has lower concentrations of solutes *[1]*; blood plasma is contained within capillaries *[1] (max 3)*

2 Lymph has less oxygen and nutrients (on average) *[1]* because these particles are taken up by cells prior to fluid draining into the lymphatic system *[1]*; lymph has more leucocytes (particularly T lymphocytes) *[1]*, which are added to the lymphatic system (once they mature in the thymus) *[1]*

3 Capillaries are permeable *[1]* because of thin, single layer of cells/gaps between cells in the wall *[1]*; oncotic/osmotic pressure is always higher in the capillary blood than the tissue fluid (or water potential is higher in the tissue fluid) *[1]*; hydrostatic pressure is high in capillary blood at the arterial end of the capillary *[1]*; hydrostatic pressure decreases towards the venous end of the capillary *[1]*; water diffuses across the capillary wall *[1]* because oncotic pressure outweighs hydrostatic pressure *(max 5)*

8.4

Go further

1 Crocodile haemoglobin contains stronger bonds in its tertiary structure to prevent structural change over a wider range of temperatures

2 Myoglobin is an oxygen store in the muscles. It releases oxygen when respiration rate in muscles is high

Myoglobin's oxygen dissociation curve is to the left of the curve for haemoglobin until approximately 9 kPa O_2 partial pressure

3 Haemocyanin lacks the iron-containing haem prosthetic group, which is responsible for the red colour of haemoglobin. (Haemocyanin instead contains copper)

Summary questions

1 a Haemoglobin's affinity for oxygen decreases as concentration (partial pressure) of CO_2 increases *[1]*

   b y-axis = % haemoglobin saturation *[1]*; x-axis = oxygen partial pressure *[1]*; the curve for the higher CO_2 partial pressure is a similar shape but to the right and below the curve for the lower partial pressure

2 Approximately 0.2   *[1]*

3 Sigmoidal/s-shaped *[1]*; haemoglobin gains oxygen in the alveoli (at high partial pressures) *[1]*; oxygen is released at respiring tissues (at low partial pressures) *[1]*

8.5

1 a Relaxation of the heart *[1]*

   b Contraction of the atria *[1]*

   c Contraction of the ventricles *[1]*

2 The electrical impulse from the SAN (P wave) reaches the AVN *[1]*; the delay is because of the conduction of the impulse down the bundle of His (before ventricular contraction (QRS)) *[1]*

3 Atrial pressure decreases during ventricular systole *[1]*; ventricular pressure is highest during ventricular systole *[1]* and lowest during diastole *[1]*

9.1

1 Central in roots;   [1]

   Towards the outside of stems   [1]

2 Lignin strengthens the vessels;   [1]

   No end wall, which enables a smooth flow of water;   [1]

   Pits enable water movement between vessels   [1]

3 Companion cells carry out the functions that the sieve tube elements cannot;   [1]

   Communication between the two types of cell occurs through plasmodesmata   [1]

9.2

1 Maintaining turgor;   [1]

   For photosynthesis;   [1]

   Transport medium;   [1]

   Cooling plants via evaporation   [1 (max 2)]

2 Forces water back into cytoplasm/symplast pathway;   [1]

   Because the strip is waterproof;   [1]

   Prevents water returning from xylem to the cortex   [1]

3 Apoplast: water moves through cell walls;   [1]

   and intercellular spaces;   [1]

   Symplast: water moves through cytoplasm;   [1]

   and plasmodesmata;   [1]

   Water experiences little resistance in the apoplast pathway   [1]

9.3

1 a Lower rate;   [1]

      Reduced kinetic energy of water molecules   [1]

   b Higher rate;   [1]

      More stomata open   [1]

   c Higher rate;   [1]

      Leaves are more permeable   [1]

2 Stomata required for gas exchange; *[1]*

But water is lost when stomata are open; *[1]*

3 Radius of tube = 0.5 mm; *[1]*

$\pi r^2 = 0.785$; *[1]*

Volume per minute = 9.42 mm³; *[1]*

Volume per hour = 565.2 mm³

[1]

9.4

1 *Sources* *(2 max)*

Green leaves; *[1]*

Green stems; *[1]*

Tubers; *[1]*

Tap roots; *[1]*

Food stores in seeds *[1]*

Sinks *(2 max)*

(Growing) roots; *[1]*

Meristems; *[1]*

Developing seeds; *[1]*

Developing fruits; *[1]*

Storage organs *[1]*

2 H⁺ ions moved out of companion cells via active transport; *[1]*

H⁺ concentration gradient established; *[1]*

Co⁻transport; *[1]*

Of H⁺ and sucrose into companion cells *[1]*

3 Glucose levels are generally lower than other sugars in all tissues; *[1]*

Glucose is converted to sucrose for transport; *[1]*

Relatively high glucose levels around sources in leaves because some glucose is yet to be converted; *[1]*

Sucrose is generally higher than other carbohydrates; *[1]*

Except in tissues surrounding vascular bundles because some of the conversion is yet to occur; *[1]*

Sucrose high in buds, roots, and tubers because it is transported to these sinks; *[1]*

Fructose is low in the leaf blades and sinks because it is not used directly by this plant; *[1]*

Fructose is relatively high in and around vascular tissue, suggesting it can be transported easily *[1 (max 6)]*

9.5

1 Reduced number of stomata; *[1]*

Reduced leaf area; *[1]*

Sunken stomata; *[1]*

Curled leaves; *[1]*

Thick cuticle; *[1]*

Increased water storage; *[1]*

Leaf loss/dormancy; *[1]*

Long roots *[1 (max 3)]*

2 Stomata on the upper surface of leaves (in hydrophytes); *[1]*

The lower surface is often not in contact with the air for gas exchange *[1]*

3 Lab conditions are different from those in the wild; *[1]*

The plant might be damaged when cut; *[1]*

An isolated shoot is smaller than the whole plant *[1]*

10.1

1 a The genus name is not capitalised *[1]*

b The binomial name is neither underlined nor italicised *[1]*

c The species name is capitalised *[1]*

2

Taxon	Human	Chimpanzee
Domain	Eukaryota	Eukaryota
Kingdom	Animalia	Animalia
Phylum	Chordata	Chordata
Class	Mammalia	Mammalia
Order	Primates	Primates
Family	Hominidae	Hominidae
Genus	*Homo*	*Pan*
Species	*sapiens*	*troglodytes*

(1 mark per row)

3 Some species reproduce asexually *[1]*; two groups of organisms can be classified as separate species yet produce fertile offspring (e.g. chimpanzees and bonobos), which suggests other criteria can be used to determine species *[1]*

10.2

1 Bacteria have peptidoglycan in their cell walls/fungi have chitin in their cell walls *[1]*; bacteria are always unicellular, whereas some fungi are multicellular *[1]*; bacteria lack organelles *[1]*; fungi can feed using saprotrophic nutrition *[1]*; fungi have hyphae *[1]* *(max 2)*

2 a Kingdom = Protoctista, Domain = Eukarya *[1]*;

b Kingdom = Fungi, Domain = Eukarya *[1]*;

c Kingdom = Prokaryotae, Domain = Archaea *[1]*

3 New evidence shows different relationships *[1]*; example of new evidence (e.g. DNA sequencing) *[1]*; two examples of the evidence used in domain classification (e.g. RNA polymerase, cell wall structure, rRNA) *[2]*

10.3

1 New evidence (such as comparative genetics) *[1]*; the evolutionary relationships between species are re-evaluated *[1]*

2 a *Homo erectus [1]*
   b Humans *[1]*

3 Evolutionary positions can be compared/ evolutionary histories can be mapped *[1]*; phylogenetic trees avoid arbitrary groupings of species *[1]*; the use of genetic and evolutionary comparisons enable more precise classification *[1]*

10.4

1 The greater the similarities in the base sequences *[1]* the closer the evolutionary relationship *[1]*

2 Gaps/missing links in the record *[1]*; some species do not fossilise (e.g. due to the lack of a skeleton) *[1]*; some fossils have been destroyed since formation *[1]*

3 Variation exists (between individuals within populations) *[1]*; both artificial selection and natural selection involve individuals with favoured traits breeding *[1]*; in both cases, the favoured traits are inherited by offspring *[1]*; over many generations allele frequencies change within the populations *[1] (3 max)*

10.5

1 a Intraspecific because the variation is within one species *[1]*

   b Interspecific because the variation is between two species *[1]*

   c Intraspecific because similar variation is shown within both species (and no information is provided about how the two species may differ) *[1]*

2 Mutation *[1]* changes DNA base sequence *[1]*; crossing over *[1]* and independent assortment (of homologous chromosomes and chromatids) *[1]* produce new combinations of alleles *[1]*; random fertilisation/mating (increases the number of possible allele combinations) *[1] (4 max)*

3 Identical twins have 100% of the same DNA and fraternal twins have approximately 50% of the same DNA *[1]*; there is a greater genetic influence on height than the risk of strokes *[1]*; the evidence for this is that more than 90% of identical twins have the same height, whereas approximately 15% of identical twins share the risk of having a stroke *[1]*

10.6

1 a Continuous *[1]*
   b Discontinuous *[1]*
   c Continuous *[1]*

2 $(\bar{x}_1 - \bar{x}_2) = 1.6$ *[1]*; $\sqrt{\left(\dfrac{\sigma_1^2}{n_1}\right) + \left(\dfrac{\sigma_2^2}{n_2}\right)} = 0.393$ *[1]*; $t = 4.068$ *[1]*; significant difference *[1]*; because p is below 0.05 *[1]*

3 (Note: some calculations for the horn length data are given in the worked example earlier in the Topic). Correctly ranked horn length and number of matings *[1]*; correct rank order differences *[1]*; $6\Sigma d^2 = 90$ *[1]*; $n(n^2 - 1) = 990$ *[1]*, $r_s = 0.909$ *[1]*; the correlation is significant *[1]*

10.7

1 a Physiological *[1]*
   b Behavioural *[1]*
   c Anatomical *[1]*
   d Physiological *[1]*

2 The two species possess similar traits *[1]* as adaptations to similar ecological niches *[1]* but the adaptations evolved independently/at different times *[1]*

3 **Level 3 *[5–6 marks]*:** Detailed description of all three types of adaptation, with relevant examples. There is a well-developed line of reasoning which is clear and logically structured. The information presented is relevant and substantiated.

   Level 2 *[3–4 marks]*: Includes at least two types of adaptation. There is a line of reasoning presented with some structure. The information presented is in the most part relevant and supported by some evidence

   Level 1 *[1–2 marks]*: Simple comments about at least one type of adaptation. The information is basic and communicated in an unstructured way. The information is supported by limited evidence.

   Relevant scientific points include:

   Anatomical: flagella for locomotion; pili for conjugation/exchange of genetic material

   Behavioural: chemotaxis

   Physiological: antibiotic resistance; receptors for binding to cells; relevant enzymes

10.8

Go further

1 The frequency of the lighter fur trait has increased in the population so the allele must have conveyed an advantage. The pale-and dark-coloured mice may still be able to breed and are therefore not separate species.

2 The survival advantage per generation is small, but the cumulative effect over many generations would have resulted in a significant shift in allele frequencies.

Summary questions

1 pH changes [1]; temperature changes [1]; a lack of water [1]; antibiotics [1]; destruction by cells in their host's immune system [1] (max 3)

2 (For example) nylonase production in *Flavobacterium* [1]; this enables industrial waste to be removed [1]

3 The insect species has variation between its members [1]; the insecticide acts as a selection pressure [1]; only individuals that have some resistance to the insecticide will survive, reproduce and pass on their beneficial alleles to their offspring [1]; over generations, the proportion of alleles for resistance will increase and the entire population will be resistant to the insecticide [1]

11.1

1 *Population*: a group of organisms of the same species; [1]

   Community: a set of populations within an ecosystem [1]

2 Species biodiversity; [1]

   Because the variety and abundance of species within the ecosystem indicates the stability and health of food chains [1]

3 Species richness might be high; [1]

   But a few species might dominate/many species have low numbers; [1]

   Biodiversity is based on both species richness and evenness [1]

11.2

1 Time; [1]

   Money; [1]

   Labour availability; [1]

   Access (to the ecosystem); [1]

   Equipment [1 (max 3)]

2 Belt transect; [1]

   Line placed along the ecosystem (from shoreline to the edge of sand dunes); [1]

   Quadrats used at intervals; [1]

   Abundance of each species estimated (e.g. by percentage cover) [1]

3 Number of samples should be maximised; [1]

   But will depend on time/resources/number of surveyors; [1]

   Stratified random sampling; [1]

   The number of samples in each of the three areas should be proportional to their size (e.g. 100 samples taken in total, 70 on grassland, 20 in the shrubby area, 10 in the orchard); [1]

Random sampling is used within each area (e.g. use of a random number generator to select coordinates); [1]

Quadrats used to estimate percentage cover of each species [1]

11.3

1 To prevent rainwater from entering and drowning the trapped organisms [1]

2 Time limitations; [1]

   In cases where individuals of a species are difficult to count [1]

3 22 (6 × 60 × 60, divided by 1000) [2]

   (one mark for 21.6 not rounded to two significant figures)

11.4

1 *n*: the number of individuals of one species; [1]

   N: the total number of individuals of all species [1]

2 Few species/low numbers in many of the species; [1]

   An environmental change may cause the loss of species from the ecosystem; [1]

   Food chains would be disrupted [1]

3 Habitat A = 0.500; [1]

   Habitat B = 0.759; [1]

   Habitat A has less biodiversity and will be more sensitive to environmental change [1]

11.5

1 Small population; [1]

   Due to disease; [1]

   Or habitat destruction; [1]

   Or migration (founder effect); [1]

   Natural selection [1 (max 3)]

2 *Idea that* high genetic diversity indicates a species that has a wide range of traits on which natural selection can act; [1]

   Idea that when a selection pressure is exerted there is a greater chance that some members of the species will adapt and survive [1]

3

Species	Number of monomorphic genes studied	Number of polymorphic genes studied	Percentage of polymorphic genes (%)	Average heterozygosity (%)	
A	28	8	22	4.2	[1]
B	35	10	22	7.0	[1]
C	8	3	27	5.4	[1]

It is not possible to conclude which species has the highest genetic diversity/the data lead to conflicting conclusions; [1]

Average heterozygosity suggests species B, % of polymorphic genes suggests species C; [1]

However, very few genes have been analysed to determine % of polymorphic genes, especially for species C. Analysing more genes would improve the validity of the results; [1]

It is unclear how many genes have been analysed to calculate average heterozygosity. This could be the more accurate of the two measures in this case [1 (max 5)]

11.6

1 Desertification; [1]

Melting frozen habitats; [1]

Changing sea currents/temperature [1 (max 2)]

2 (Reduced biodiversity due to) eutrophication; [1]

Toxic chemicals poisoning aquatic organisms; [1]

Acid rain production altering pH in aquatic ecosystems; [1]

Global warming altering sea currents/ temperature; [1]

Drainage of rivers and lakes [1 (max 4)]

3 Intensive farming increases crop yields; [1]

But requires more labour; [1]

And more fertilisers/pesticides/herbicides; [1]

And exhibits lower animal welfare standards than extensive farming; [1]

Intensive farming is more likely to damage ecosystems [1 (max 3)]

11.7

1 A greater chance of future drug/genetic resource discoveries; [1]

Ecotourism; [1]

A greater gene pool for artificial selection [1 (max 2)]

2 Reduces the chances of soil degradation; [1]

Land remains viable for longer; [1]

Biodiversity is maintained; [1]

Energy/food requirements can be met for a longer period of time [1 (max 2)]

3 Otters prevent an overpopulation of urchins; [1]

Which would severely reduce the kelp population [1]

11.8

1 Sustainability targets; [1]

Targets to reduce desertification/increase land fertility; [1]

Targets for greenhouse gas reductions [1 (max 2)]

2 Seeds are smaller than plants, therefore easier to transport; [1]

And less space is required for storage; [1]

Lower probability of disease/damage; [1]

Long-term storage/viability maintained for many years [1 (max 3)]

3 *In situ* is (usually) cheaper; [1]

In situ enables interspecies relationships to be maintained; [1]

In situ avoids potential problems when reintroducing species; [1]

Ex situ provides optimum conditions/ veterinary care; [1]

Ex situ enables controlled breeding programmes [1]

12.1

1 Protoctista; [1]

Fungi [1]

2 a Toxin excretion by bacteria [1]

b Enzyme secretion by fungi, causing host tissue to be digested [1]

3 Viruses can reproduce; [1]

But not without exploiting the metabolism of host cells; [1]

They cannot synthesise proteins or transform energy; [1]

They have evolved over time [1 (max 3)]

12.2

Go further

No reliable cure/treatment;

No vaccine;

Further evolution (e.g. if the pathogen evolves airborne transmission) would increase infection rates;

Correct identification of symptoms might be difficult

Summary questions

1 *Similarity*: both involve hyphae penetrating plant tissue; [1]

Difference: potato blight is caused by a protoctist, whereas black sigatoka is caused by a fungus; [1]

Black sigatoka infects only leaves [1]

2 In T helper cells; [1]

Reverse transcriptase; [1]

Converts viral RNA into DNA; [1]

Viral DNA integrated into host cell DNA; [1]

Viral proteins and RNA replicated using host machinery [1 (max 4)]

3 Genetic mutation resulted in new antigens on the H1N1 virus; [1]

New strain not encountered before by human immune systems; [1]

No vaccine *[1 (max 2)]*

12.3

1 High population density/overcrowding; [1]

Poor nutrition; [1]

Poor hygiene/waste disposal; [1]

Culture/medical practices; [1]

Number of trained health professionals; [1]

Lack of public warning systems *[1 (max 4)]*

2 An inanimate object that can harbour and spread pathogens; [1]

Examples include:

Meningitis; [1]

Influenza; [1]

Athlete's foot; [1]

Cold sores; [1]

Conjunctivitis *[1 (max 2 for examples)]*

3 Soil contamination; [1]

With bacteria [1]

12.4

1 (Named) antibacterial compound(s); [1]

(Named) antifungal compound(s); [1]

Anti-oomycetes/glucanases; [1]

Cyanide compounds *[1 (max 2)]*

(*Note: insect repellents and insecticides should not be accepted as answers because insects are pests rather than pathogens*)

2 a To provide a physical barrier against pathogens [1]

b To stop pathogens moving through plasmodesmata to infect neighbouring cells [1]

c To stop pathogens moving through phloem sieve tubes to other parts of the plant [1]

3 Both are formed from β-glucose monomers; [1]

Callose has 1,3 glycosidic bonds/cellulose has 1,4 glycosidic bonds; [1]

Both are (largely) linear (although callose has some branches); [1]

Callose is helical/cellulose is not helical; [1]

Cellulose has cross-links between chains *[1 (max 4)]*

12.5

1 They defend against any pathogen [1]

2 *Thrombin*: catalyses the conversion of fibrinogen to fibrin; [1]

Thromboplastin: catalyses the conversion of prothrombin to thrombin [1]

3 Mast cells release histamines; [1]

Body temperature is raised, which inhibits pathogen reproduction; [1]

And causes inflammation/swelling [1]

12.6

1 Perforin production; [1]

Disruption of pathogen cell membranes [1]

2 Antibody (complex) with many binding sites for antigen; [1]

Pathogens clumped together; [1]

Complex is too large to enter cells; [1]

Facilitates phagocytosis/phagocytes can engulf several pathogens at once *[1 (max 3)]*

3 T helper lymphocytes infected/destroyed; [1]

Cytokines not produced to activate specific B memory cells [1]

12.7

1 Antibodies are injected into a person; [1]

No memory cells are developed [1]

2 Many people in a population have immunity to a pathogen; [1]

A disease-carrier is less likely to encounter people who lack immunity [1]

3 The tetanus vaccine is in the form of toxoids/modified toxins; [1]

Small amounts of toxin can be very harmful; [1]

Boosters maintain high levels of antitoxin antibodies and memory cells in the blood (to maintain immunity) [1]

13.1

Go further: Types of cell signalling

1 a Paracrine signalling; the presynaptic and postsynaptic neurones are close to each other, separated only by the synaptic cleft. Neurotransmitters diffuse between the two neurones.

b Endocrine signalling; insulin (a hormone) is released from beta cells and passes through the blood to target cells in the liver.

Summary questions

1 Neurotransmitters; *[1]*

Postsynaptic; *[1]*

Endocrine *[1]*

2 Transmits nervous impulses between neurones; *[1]*

Across synapses / synaptic clefts; *[1]*

For coordinated responses to stimuli; *[1]*

idea of summation *[1][2 max]*

3 Receptor cells need to communicate with controller / coordinator / brain; *[1]*

Controller / coordinator / brain needs to communicate with effectors; *[1]*

idea of several cell types / tissues / systems involved in responses *[1] [2 max]*

13.2

1 Position of cell body (described); *[1]*

Sensory neurone has one (long) dendron; *[1]*

Motor neurone has many dendrites leading to its cell body *[1] [2 max; reverse arguments apply]*

2 Relay neurones tend to be short; *[1]*

The increased rate of conduction in the presence of myelin sheath would make little difference *[1]*

3 a Proteins are synthesised in the cell body; *[1]*

Some proteins must travel to other regions of the neurone (e.g. motor junctions at the other end of the cell); *[1]*

b Vesicle (production) / cytoskeleton / motor proteins *[1]*

13.3

1 (Sensory receptors) convert one form of energy to another form of energy *[1]*

2 a Light; *[1]*

b Heat / thermal; *[1]*

c Mechanical pressure; *[1]*

d Chemical *[1]*

3 *In response to hormones*:

cAMP is a second messenger; *[1]*

stimulates cascade of reactions (in target cell) *[1]*

In sensory receptor:

Alters cell membrane permeability; *[1]*

To sodium ions; *[1]*

Stimulates depolarisation / action potentials *[1] [3 max]*

13.4

1 three; *[1]*

two; *[1]*

depolarised; *[1]*

repolarisation *[1]*

2 *(with myelination)*

Faster impulses; *[1]*

Saltatory conduction; *[1]*

Longer local circuits; *[1]*

Nodes of Ranvier present
[1] [3 max] (allow reverse argument throughout)

3 All-or-nothing principle; *[1]*

Threshold potential must be surpassed; *[1]*

No action potential will occur if Na^+ ion influx is insufficient; *[1]*

Stimulus strength does not affect action potential magnitude (above the threshold); *[1]*

K^+ ion channels always open at the same point in an action potential (which results in the potential difference never surpassing a particular value)
[1] [3 max]

13.5

1 Calcium ion channels open / calcium ions diffuse into the presynaptic neurone; *[1]*

Vesicles fuse with the presynaptic membrane; *[1]*

Acetylcholine is released by exocytosis *[1]*

2 A weak stimulus produces less frequent action potentials; *[1]*

Fewer neurotransmitter molecules are released; *[1]*

Less depolarisation (in postsynaptic neurone); *[1]*

Threshold is not reached; *[1]*

No postsynaptic action potential *[1] [3 max]*

3 To be released again from the presynaptic neurone; *[1]*

Regulates the concentration of neurotransmitter in the synaptic cleft; *[1]*

Prevents re-binding to receptors when signalling has stopped *[1] [2 max]*

13.6/13.7

1 a somatic; *[1]*

   b sympathetic; *[1]*

   c parasympathetic *[1]*

2 The cerebrum controls voluntary movements; *[1]*

   The cerebellum coordinates balance and non-voluntary movements *[1]*

3 Humans have a larger brain than expected for their body mass; *[1]*

   This may represent an evolutionary adaptation for problem solving / higher thought processes / language / complex learning / social interactions *[1]*

13.8

1 a spinal cord; *[1]*

   b brain stem *[1]*

2 Few synapses; *[1]*

   Short pathway; *[1]*

   A dangerous stimulus is responded to quickly; *[1]*

   Innate /no learning required; *[1]*

   The response is stereotyped/consistent/always the same; *[1]*

   Involuntary; *[1]*

   Prevents overloading of the brain *[1] [max 4]*

3 Two synapses; *[1]*

   Short relay neurone; *[1]*

   Sensory neurone does not need to travel deep into the brain; *[1]*

   Response is quicker *[1] [max 3]*

13.9/13.10

1 a no change; *[1]*

   b shortens; *[1]*

   c no change; *[1]*

   d no change; *[1]*

   e shortens *[1]*

2 Calcium ions bind to troponin; *[1]*

   Changes the shape of troponin; *[1]*

   Displaces tropomyosin; *[1]*

   Uncovers myosin binding sites on actin; *[1]*

   Enables myosin heads to bind to actin *[1] [max 3]*

3 *(Muscle fibres have)*

   More mitochondria; *[1]*

   To generate ATP in aerobic respiration *[1]*

   More ribosomes; *[1]*

   For actin/myosin/troponin/tropomyosin production *[1]*

Sarcoplasmic reticulum; *[1]*

To store and release calcium ions *[1]*

Myofibrils; *[1]*

To act as contractile organelles *[1]*

Many nuclei; *[1]*

Due to the fusion of muscle cells
   [1] [max 6] (accept reverse arguments when relevant)

14.1

1 a The cell to which a hormone binds and in which it produces an effect; *[1]*

   b A substance that is activated by a hormone; *[1]*

      And produces a response within a target cell *[1]*

2 *Similarities:*

   transported in blood; *[1]*

   produce responses in target cells; *[1]*

   bind to receptors; *[1]*

   regulate transcription *[1] [max 2]*

   Differences:

   steroid hormones diffuse through cell membranes (non-steroid hormones do not); *[1]*

   steroid hormone receptors are in the cytoplasm or nucleus / non-steroid hormone receptors are on the cell surface membrane; *[1]*

   non-steroid hormones activate second messengers / reaction cascades *[1] [max 2]*

3 Glucocorticoids secreted, which regulate carbohydrate metabolism; *[1]*

   More (named) respiratory substrates made available; *[1]*

   Mineralocorticoids (e.g. aldosterone) enable more water to be reabsorbed from the kidneys; *[1]*

   Adrenaline can raise heart rate (during the race); *[1]*

   Adrenaline promotes glycogenolysis / raises blood glucose concentration; *[1]*

   Noradrenaline increases heart rate / widens airways *[1] [max 5]*

14.2

1 Differential staining; *[1]*

   Islets of Langerhans stain blue/lilac and acini stain dark pink/purple; *[1]*

   Acini cells are arranged in small clusters / cells in the islets are arranged in larger clusters *[1] [max 2]*

2 Endocrine glands secrete hormones into the blood; *[1]*

exocrine glands secrete other substances such
as enzymes; [1]

the pancreas operates as both types of gland; [1]

it secretes digestive enzymes; [1]

it secretes the hormones insulin and glucagon [1] [max 4]

3 β cells have potassium ion channels that close
when glucose concentrations are high; [1]

β cells have calcium ion channels that open
when glucose concentrations are high; [1]

β cells transport insulin by exocytosis; [1]

β cells have glucose receptors/transporters; [1]

α cells transport glucagon by exocytosis; [1]

α cells (also) have glucose receptors/transporters; [1]

idea that α cells secrete glucagon when little
glucose is entering the cells [1] [max 5]

14.3

1 (ATP is) produced when glucose enters the cell; [1]

Binds to potassium ion channels; [1]

Which causes depolarisation / opens calcium
ion channels [1] [max 2]

2 9×10^{-4} (or 0.0009) [1]

3 Liver cells contain stores of glycogen; [1]

glucagon is secreted by the pancreas in response
to low blood glucose concentration; [1]

when glucagon binds to receptors on liver cells,
glycogen within these cells is broken down into
glucose molecules [1]

14.4

Go further: Maturity-onset diabetes of the young

1 The condition is inherited via a single gene, whereas
type 1 and type 2 diabetes appear to have more
complicated genetic influences (e.g. it is likely that
many different genetic variants increase a person's
susceptibility for developing type 2 diabetes).

2 0.75–1.00 / 75–100% (0.75 if both parents
are heterozygous and 1.00 if either parent is
homozygous dominant for the MODY allele).

3 The homozygous genotype (i.e. having two copies
of the defective MODY gene variant) produces
symptoms that are more severe.

Summary questions

1 Type 1 is insulin-dependent because people with
this form produce insufficient insulin; [1]

type 2 is insulin-independent because people with
this form often produce sufficient insulin but the
target cells are insensitive to the hormone [1]

2 Blood glucose increases to a greater extent in
the diabetic; [1]

blood glucose takes longer to return to the original
level in the diabetic; [1]

the diabetic produces no additional insulin because
their pancreatic beta cells have been destroyed [1]

3 Type 1 diabetes is not affected by lifestyle
to a great extent; [1]

the prevalence of type 1 diabetes is unlikely to
change much; [1]

the prevalence of type 2 diabetes is increasing; [1]

this is because of an aging population and an
increase in obesity levels; [1]

the increasing prevalence will cost health care
services more money to treat and manage in
the future. [1]

14.5

1 (Adenylyl cyclase is) activated by (the binding of)
adrenaline; [1]

on liver cell surface membrane; [1]

(Adenylyl cyclase) converts ATP; [1]

into cAMP [1] [max 2]

2 Both systems work together to produce
responses; [1]

The sympathetic nervous system stimulates the
release of hormones [1]

3 Hearing is not (usually) essential for responding
to threats; [1]

Blood flow is diverted from the auditory region of
the brain to other regions [1]

14.6

1 Chemoreceptors detect changes in blood pH that
reflect carbon dioxide levels; [1]

pressure receptors detect changes in
blood pressure; [1]

the receptors pass information to the medulla
oblongata, which initiates the necessary response [1]

2 Increased physical activity raises the amount of
energy required by muscles; [1]

muscles require more oxygen and glucose for
respiration; [1]

cardiac output must be increased to supply muscles
with these molecules at the necessary rate [1]

3 Blood pressure would remain high; [1]

the parasympathetic nervous system is no longer
able to carry impulses / transmit to the heart; [1]

the medulla oblongata cannot stimulate the SAN to
lower heart rate [1]

15.1

1 a The desired value around which negative feedback operates.   *[1]*

   b The range of values across which a physiological factor varies and negative feedback operates.   *[1]*

2 Oxytocin increases the strength and frequency of contractions;   *[1]*

   the increase in contractions causes more oxytocin to be released;   *[1]*

   contractions are therefore intensified further and gradually increased;   *[1]*

   the gradual increase in contractions enables a baby to be born using the minimum intensity of contraction.   *[1]*

3 Homeostasis maintains conditions within a narrow range around an optimum value (set point);   *[1]*

   negative feedback reverses any deviation away from the set point;   *[1]*

   positive feedback increases any change made to a physiological factor.   *[1]*

15.2/15.3

1 Arterioles (and shunt vessels) ;   *[1]*

   Sweat glands ;   *[1]*

   Hair erector muscles ;   *[1]*

   Fat tissue   *[1]*

2 *Advantages*

   Lower food requirements (than endotherms);   *[1]*

   A greater proportion of energy intake can be used for growth   *[1]*

   Disadvantages

   Lower activity levels in cold temperatures (compared to endotherms);   *[1]*

   (Therefore) greater risk of predation;   *[1]*

   May need to survive winter without food intake (due to lack of activity)   *[1] [max 4]*

3 More heat is lost to the environment;   *[1]*

   More food provides additional substrates for respiration;   *[1]*

   More respiration increases metabolic rate;   *[1]*

   More heat is generated (in response to the colder winter temperatures);   *[1]*

   Some species, especially hibernating species, increase their fat reserves   *[1] [max 3]*

15.4

1 Hydrogen peroxide;   *[1]*

   ethanol;   *[1]*

   a (named) drug   *[1] [max 2]*

2 (Transamination is) the conversion of one amino acid into another;   *[1]*

   some amino acids are not supplied in an organism's diet   *[1]*

3 In hepatocytes / intracellular enzymes;   *[1]*

   enzymes control deamination and the reactions of the ornithine cycle;   *[1]*

   (glycogen synthase) catalyses glucose to glycogen conversion;   *[1]*

   (glycogen phosphorylase) catalyses glycogen to glucose conversion;   *[1]*

   Catalase;   *[1]*

   breaks down hydrogen peroxide ;   *[1]*

   Alcohol dehydrogenase ;   *[1]*

   removes ethanol   *[1] [max 5]*

15.5

Go further: Diuretics – treating, cheating, and energy-depleting?

1 Fewer sodium and chloride ions are co-transported from the distal convoluted tubule into the tissue fluid; the DCT water potential remains lower than it otherwise would be; less water diffuses out of the nephron by osmosis.

2 Alcohol inhibits ADH, which causes a greater volume of urine to be produced; drinking water replaces some of the lost fluid; this helps to lower the solute concentration of blood.

Summary questions

1 The renal vein will have a lower concentration of urea (and other toxins);   *[1]*

   water potential may be different in the two blood vessels   *[1]*

2 To transport (the few) proteins that have been filtered into the nephron;   *[1]*

   endocytosis transports proteins from the PCT lumen into cells in its wall;   *[1]*

   exocytosis transports proteins from cells in the PCT wall into tissue fluid   *[1]*

3 Dialysis does not provide a cure, only an improvement in condition for the patient;   *[1]*

   successful kidney transplantation provides a cure; *[1]*

   however, transplant surgery carries a high risk of organ rejection;   *[1]*

   future transplant surgery may use therapeutic cloning to reduce the possibility of rejection;   *[1]*

   haemodialysis usually requires regular trips to a health clinic and is time-consuming;   *[1]*

   peritoneal dialysis can be carried out by the patient while they work, but must be done every day   *[1]*

16.1

1   Cell elongation;   [1]

    apical dominance / shoot growth;   [1]

    root growth;   [1]

    (photo)tropism(s);   [1]

    ethene release / fruit ripening;   [1]

    prevention of abscission / leaf fall   [1] [max 3]

2   *Similarities [max 2]*

    Cell signalling molecules;   [1]

    receptors;   [1]

    idea of often work as antagonists/in opposition
    (e.g. glucagon and insulin, gibberellin and ABA)   [1]

    Differences [max 2]

    Animals have endocrine glands / plants
    lack glands;   [1]

    animal hormones move in blood and plant
    hormones move in phloem/xylem/through cells;   [1]

    animal hormones produce effects more rapidly   [1]

3   0.005 (5×10^{-3}) mol dm^{-3} [1]

    $0.000\,05$ (5×10^{-5}) moles   [1]

16.2/16.3

Go further: Flowering – the long and the short of it

1   Long-day plants: The proportion of P_{fr} increases as
    day length increases and night length decreases,
    which initiates flowering.

    Short-day plants: The proportion of P_r increases as
    day length decreases and night length increases,
    which initiates flowering.

Summary questions

1   Physical defences;   [1]

    thorns/barbs/spikes/stings;   [1]

    fibrous/inedible tissue;   [1]

    chemical defences;   [1]

    (named) toxic compound;   [1]

    pheromones for communication between
    plants   [1] [max 4]

2   Auxin concentration reduced;   [1]

    increased ethene sensitivity (in abscission zone); [1]

    cellulase/digestive enzyme genes switched on/
    transcribed;   [1]

    increased cellulase production;   [1]

    cell walls digested in abscission zone   [1][max 4]

3   Pheromones enable communication
    between plants;   [1]

    neighbouring plant produces callose (before
    infection)   [1]

16.4

1   Plant is grown on its side;   [1]

    constant rotation;   [1]

    idea that gravitational force is felt evenly by all
    parts of the plant;   [1]

    roots and shoots grow straight / horizontally
                                                [1] [2 max]

2   Transport of auxin from tip/apex down the shoot;   [1]

    protons/H$^+$ ions pumped into cell walls;   [1]

    causing cell elongation   [1]

3   Positive phototropism increases photosynthetic
    rate;   [1]

    more carbohydrates produced;   [1]

    idea that negative phototropism causes roots to
    grow further into soil;   [1]

    more water/minerals absorbed   [1]

16.5

1   Promotes/speeds up fruit dropping at high
    concentrations;   [1]

    slows down/prevents fruit dropping at low
    concentrations;   [1]

    slows down leaf fall;   [1]

    promotes root growth;   [1]

    weedkillers   [1] [3 max]

2   Ethene cannot be sprayed in liquid form;   [1]

    gas release is less controlled   [1]

3   Increases/speeds up ripening [no mark]

    Explanation:

    Rise in carbon dioxide indicates an increase in
    respiration rate;   [1]

    which indicates polysaccharides are being broken
    down to glucose/sucrose/maltose/fructose/
    respiratory substrates (during the ripening process)
                                                        [1]

17.1

1   Endothermic;   [1]

    active;   [1]

    ATP   [1]

2   Use of electron transport chains;   [1]

    use of coenzymes (e.g. NAD in respiration and
    NADP in photosynthesis);   [1]

    some of the same intermediates are formed in
    both (e.g. GP);   [1]

    starting materials are regenerated in both;   [1]

    both use chemiosmosis;   [1]

    to produce ATP   [1] [max 3]

3 Light energy (photons) converted to chemical energy in organic molecules in photosynthesis; *[1]*

respiration converts chemical energy in organic molecules into chemical energy in inorganic molecules (ATP); *[1]*

movement of H^+ ions (kinetic energy) drives ATP production via chemiosmosis; *[1]*

chemical energy is converted to thermal energy via metabolic reactions *[1] [max 3]*

17.2

1 Excited electrons have different origins (i.e. light absorption in photosynthesis, and from organic reactions in respiration); *[1]*

locations differ (i.e. thylakoid membranes in photosynthesis, and inner mitochondrial membranes in respiration) *[1]*

2 Excited electrons are passed along electron transport chains; *[1]*

energy is released as the electrons are passed to lower energy levels; *[1]*

the energy is used to pump protons; *[1]*

from the stroma to the thylakoid space/lumen *[1] [max 3]*

3 a The intermembrane space has a greater concentration of H^+ ions; *[1]*

this suggests H^+ ions have been transported into the space from the matrix *[1]*

b The positive charge of the protons in the intermembrane space creates the potential difference *[1]*

17.3

Go further: A visit to the GP

1 a ATP

b NADPH (reduced NADP)

2 CH_2OH

3 NO_3^- (nitrate ions) would be required to provide N for the amine group

Summary questions

1 Glycerol; *[1]*

triglyceride/phospholipid/plasma membrane production *[1]*

Glucose/fructose; *[1]*

respiratory substrates/polysaccharide formation *[1]*

Amino acids; *[1]*

protein/enzyme formation *[1] [max 4]*

2 A = 0.750 *[1]*

B = 0.375 *[1]*

3 ATP and NADPH are both produced in the light-dependent reactions; *[1]*

Both molecules are required for reactions in the Calvin cycle *[1]*

4 102 kg *[2]*

5 Photosynthesis only occurs in the light; *[1]*

the rate of production is insufficient to supply the plant with the concentration of ATP required; *[1]*

some plant cells lack chloroplasts and would be unable to generate ATP *[1]*

17.4

1 (*Right of the line*)

RuBP accumulates because it has less CO_2 with which to react; *[1]*

GP concentration decreases because less RuBP and CO_2 are reacting together *[1]*

(*Left of the line*)

RuBP and GP are reacting at the same rate as they are being produced; *[1] [max 2]*

2 Temperature affects the rate of enzyme activity; *[1]*

more enzymes are involved in the light-independent stage than the light-dependent stage *[1]*

3 (*At low light intensity*)

GP accumulates because a lack of ATP and NADPH are produced in the light-dependent stage; *[1]*

RuBP and TP concentrations decrease because GP is not converted to TP; *[1]*

TP is not converted to RuBP; *[1]*

ATP and NADPH are required to convert GP to TP; *[1]*

ATP is required to convert TP to RuBP *[1]*

At high light intensity substances are reacting at the same rate as they are being produced *[1] [max 3]*

18.1

1 2 ATP used to convert glucose to hexose 1,6-bisphosphate; *[1]*

4 ATP produced in converting TP to pyruvate; *[1]*

Net of 2 ATP produced *[1]*

2 7 ATP per glucose; *[1]*

2 ATP from substrate-level phosphorylation; *[1]*

2 reduced NAD produced in glycolysis; *[1]*

which result in a maximum of 5 ATP; *[1]*

in oxidative phosphorylation *[1] [max 4]*

3 ATP phosphorylates glucose; *[1]*

to prepare it for subsequent reactions / to be broken down; *[1]*

TP is phosphorylated; *[1]*

by inorganic phosphate; [1]

ADP is phosphorylated when TP is
converted to pyruvate [1] [max 4]

18.2

1 Enters mitochondria; [1]

(through) active transport; [1]

converted to acetyl groups; [1]

decarboxylated / loses carbon; [1]

dehydrogenated / loses hydrogen [1] [max 4]

2 Leaves the plant through stomata; [1]

(or) used in photosynthesis [1]

3 Subsequent stages of aerobic respiration take
place in mitochondria; [1]

Enzymes of the Krebs cycle are located in
mitochondrial matrix; [1]

Electron transport chains are located on
inner mitochondrial membrane [1] [max 2]

18.3

1 4 [1]

2 10 ATP molecules (per turn); [1]

1 ATP from substrate-level phosphorylation; [1]

7.5 ATP from oxidative phosphorylation via
reduced NAD; [1]

1.5 ATP via FAD [1]

3 *Idea of* rapid turnover / constantly reacting; [1]

reacts with acetyl CoA to form citrate [1]

18.4

Go further: How many ATP molecules are produced per
glucose molecule?

1 Reduced NAD must be transported from the
cytoplasm into the matrix. This requires ATP,
and the reduced NAD may enter the ETC at a
different point.

2 Reduced NAD enters at carrier 1 (because it is
estimated to be responsible for 2.5 ATP molecules,
which = 10 H^+ ions (from the four pumps) / 4

Reduced FAD enters at carrier 2 (because it is
estimated to be responsible for 1.5 ATP molecules,
which = 6 H^+ ions (from the final two pumps) / 4

Summary questions

1 Cristae are folds of the inner membrane; [1]

which increase the surface area available for
oxidative phosphorylation [1]

2 Oxygen accepts electrons that have passed
through the ETC; [1]

forming water; [1]

when combined with H^+ ions [1]

3 Some ATP is used for active transport of pyruvate
(from the cytoplasm); [1]

ETC is inefficient / some energy is released as
heat (and is not used to transport H^+ ions); [1]

some reduced coenzymes may be used in
other processes (not the ETC) [1] [max 2]

18.5

1 Relatively small amounts of ATP are generated; [1]

insufficient chemical energy to sustain the
functions of mammalian bodies [1]

2 Aerobic = 34.0% $(((32 \times 30.6) / 2880) \times 100)$; [1]

anaerobic = 2.1% $(((2 \times 30.6) / 2880) \times 100)$ [1]

3 a *Idea of* no terminal electron acceptor to accept
electrons from the final electron carrier
protein complex; [1]

proton gradient is disrupted because oxygen
is not present to remove H^+ ions diffusing
through ATP synthase [1]

b NAD is not regenerated because the
ETC stops; [1]

the reactions needed to convert citrate back
to oxaloacetate cannot occur [1]

18.6

1 Amine group; [1]

deamination [1]

2 $C_{15}H_{31}COOH + 23O_2 \rightarrow$ [1]

$16CO_2 + 16H_2O$ [1]

RQ = 16/23 = 0.696 [1]

3 a Oxaloacetate; [1]

both have 4 C atoms [1]

b Pyruvate; [1]

both have 3 C atoms [1]

19.1

Go further: The FTO 'hunger' gene: an example
of the subtle effects of gene mutation?

1 Point mutation / substitution. The gene variants all
produce functional proteins, albeit differing in their
effect. Insertion and deletion mutations tend to have
a more severe effect on phenotypes.

2 It might regulate the transcription of the ghrelin
'hunger hormone' (i.e. influences gene expression).

Summary questions

1 (The genetic code is a) triplet code; [1]

Three nucleotides represent a codon / code for
one amino acid; [1]

Other codons in the DNA sequence will not
be altered [1]

2 a 2 [1]

b 1 [1]

c 4 [1]

3 3 substitution mutations:

A to T (10th base); [1]

A to T (12th base); [1]

C to T (22nd base); [1]

deletion of 8 nucleotides (GATTATGG) [1]

19.2

1 Introns are removed from pre-mRNA; [1]

the remaining exons are spliced together to form mature mRNA [1]

2 Alternative splicing enables many different proteins to be translated from a single gene; [1]

this increases genetic diversity, which allows greater complexity in organisms [1]

3 Identical twins have 100% of the same DNA and fraternal twins have approximately 50% of the same DNA; [1]

there is a greater genetic influence on height than the risk of strokes; [1]

the evidence for this is that more than 90% of identical twins have the same height, whereas approximately 15% of identical twins share the risk of having a stroke [1]

19.3

1 The formation of connections between neurones; [1]

destruction of harmful immune cells; [1]

forming the shapes of organs and tissues; [1]

removing excess cells (e.g. from between digits); [1]

any other valid suggestion [1] [max 2]

2 Species differ in the complexity of their anatomy; [1]

the number of homeobox genes increases with complexity [1]

3 During necrosis, cells rupture and release enzymes; [1]

necrosis is an uncontrolled process; [1]

apoptosis is a controlled, regulated process; [1]

rather than cells rupturing, cell fragments are packaged into vesicles [1]

20.1

1 a Discontinuous; [1]

b continuous; [1]

c continuous [1]

2 Genotypes / genetics are identical (in clones); [1]

environmental differences (in temperature); [1]

the fruit grown at the higher temperature will be larger; [1]

due to greater enzyme activity; [1]

fruit size shows continuous variation [1] [max 3]

3 Genetics are controlled / identical in twins; [1]

the relative influence of genetics and the environment can be analysed; [1]

by monitoring differences that develop between the twins; [1]

idea that a trait that shows little variation between twins is likely to be largely determined by genetics (or reverse argument) [1] [max 3]

20.2

1 X chromosome is larger; [1]

X chromosome contains more genes than the Y chromosome [1]

2 $C^R C^R$ [1]

$C^R C^W$ [1]

3 a 0% [1]

b 50% [1]

c 12.5% [1]

d 50% [1]

20.3

1 RR [1]

WW [1]

2 Round, yellow = 12/16 (75%) [1]

Round, green = 4/16 (25%) [1]

3 The genes are not linked / not on the same chromosome; [1]

the genes do not interact / there is no epistasis [1]

20.4

1 (Autosomal) linkage; [1]

genes on the same chromosome; [1]

allele combinations inherited together [1] [max 2]

2 Chi-squared value = 1.00; [1]

greater than 5% probability that the differences between observed and expected results are due to chance; [1]

scientist's predictions are supported / no significant difference between observed and expected phenotypes; [1]

genotypes are RrGG and rrGG (or rrGg) [1]

3 a (Epistatic gene produces) enzyme; [1]

homozygous recessive genotype results in no enzyme production; [1]

precursor molecule not converted [1] [max 2]

b (Epistatic gene produces) inhibitor / suppressor protein; *[1]*

Idea of modifies the other gene product; *[1]*

affects / AW transcription *[1] [max 2]*

20.5

1 Decrease in population size; *[1]*

some alleles are lost from the population *[1]*

2 $q = 0.008$ (0.007 669); *[1]*

$q^2 = 0.0000588$ *[1]*

3 13.1% are carriers *[1]*

$q = 0.0707$; *[1]*

$p = 0.929\ 29$; *[1]*

$2pq = 0.131\ 42$; *[1]*

0.8% of the general population are carriers, which is 12.3% less *[1]*

20.6

1 **a** Geographical isolation; *[1]*

b mechanical/anatomical isolation *[1]*

2 Sympatric speciation; *[1]*

behavioural isolation; *[1]*

gene flow restricted between the two groups *[1] [max 2]*

3 No: The genomes/DNA sequences of the two species are sufficiently different to consider them separate species; *[1]*

Yes: geographical isolation has caused their phenotypes to diverge, but not yet to the point where they are unable to breed together. *[1]*

21.1

1 In both procedures, molecules are separated by size; *[1]*

some molecules are slowed (by the stationary phase in chromatography) more than others *[1]*

2 **a** $10^{1.50}$ *[1]*

b $10^{3.61}$ *[1]*

c $10^{5.12}$ *[1]*

3 *Similarities:*

Each new DNA molecule consists of one old (template) strand and one new strand; *[1]*

free complementary nucleotides are joined to a template strand; *[1]*

Differences:

Only short fragments are replicated in PCR, whereas entire chromosomes are replicated naturally; *[1]*

PCR requires primers to be used; *[1]*

the DNA helicase enzyme separates strands in nature, whereas temperature cycling controls the process in PCR *[1] [max 4]*

21.2/21.3

Go further: Advances in sequencing

1 Cost, the length of DNA strands that can be sequenced during a single run, accuracy, the speed of sequencing.

Summary questions

1 **a** Particular allele base sequences are associated with certain diseases; *[1]*

People can be screened for the base sequences *[1]*

b Particular base sequences (barcodes) are unique to species; *[1]*

Similarities and differences in base sequences indicate the relatedness of different species *[1]*

2 *Similarities:*

Both use DNA polymerase; *[1]*

Both use free nucleotides *[1]*

Differences:

PCR for sequencing includes terminator bases; *[1]*

Different lengths of DNA are produced when sequencing *[1]*

Amplification uses a thermocycler, but sequencing uses capillaries / slides *[1] [max 4]*

3 mRNA can be modified after transcription; *[1]*

RNA splicing/ introns removed from RNA; *[1]*

Proteins can be modified after translation *[1]*

21.4

1 Restriction enzymes cut at specific recognition sites; *[1]*

the gene and the plasmid must have complementary sticky ends to be able to join together *[1]*

2 The herbicide will kill only weeds that are competing with the crop for resources; *[1]*

crop yield will increase and food prices will remain low *[1]*

3 The gene being studied is inactivated in embryonic stem cells (of mice); *[1]*

all cells in the study animal have the inactivated gene; *[1]*

example of gene that could be studied (e.g. tumour suppressor gene or proto-oncogene); *[1]*

the phenotypes of the knockout mice and normal mice are compared; *[1]*

the incidence of cancer is recorded *[1] [max 2]*

21.5

1 Somatic cell therapy is used; *[1]*

   this treats only the affected somatic tissue and the new alleles are not passed on to future generations; *[1]*

   the introduced alleles are not copied in mitosis; *[1]*

   the treated cells are replaced, meaning additional treatment is required *[1] [max 2]*

2 Delivering even a single gene via a vector is difficult; *[1]*

   delivering more than one gene would require a vector large enough to carry even more DNA; *[1]*

   the probability of all the new genes being integrated and functioning would be very low *[1] [max 2]*

3 The use of viruses as vectors taps into and adapts their natural mode of infection (i.e. inserting their DNA into specific cells); *[1]*

   liposomes are hydrophobic and are able to move through cell membranes *[1]*

22.1/22.2

1 (Meristem cells are) stem cells; *[1]*

   (Meristem cells are) totipotent; *[1]*

   (Meristem cells can) differentiate into a whole plant; *[1]*

   (by) vegetative propagation; *[1] [3 max]*

2 Auxin promotes root growth; *[1]*

   Enables differentiation of callus cells; *[1]*

   Enables plantlet formation; *[1] [2 max]*

3 Clones/new plants are genetically identical to the parent plant (i.e. no genetic variation); *[1]*

   Clones/new plants are equally susceptible to the pathogen; *[1]*

   The pathogen may be systemic/remains in the clones *[1] [2 max]*

22.3

Go further: Aphids

1 Reproduction is rapid (i.e. a faster rate than sexual reproduction would provide) and enables aphid populations to exploit beneficial conditions.

2 Eggs are more likely than nymphs to survive the cold temperatures of winter. Sexual reproduction also introduces genetic variation.

Summary questions

1 Reproductive cloning results in cloned whole organisms; *[1]*

   Non-reproductive cloning results in cloned cells/tissues/organs (but not whole organisms) *[1]*

2 Cells early in an embryo's development are totipotent; *[1]*

   These cells can develop into a whole organism; *[1]*

   Cells at a later stage are pluripotent/multipotent/not totipotent *[1][2 max]*

3 (Yes)

   No nuclear DNA will be present from the egg donor; *[1]*

   But mitochondrial DNA is transferred with the egg *[1]*

22.4/22.5

1 A process carried out by a microorganism is exploited for commercial benefit *[1]*

2 Microorganisms consume nutrients from the waste water; *[1]*

   (Organic) waste matter is therefore removed from the water *[1]*

3 Antibiotic chemicals are secreted by *Penicillium*; *[1]*

   The antibiotics are toxic to other microorganisms; *[1]*

   Food sources are preserved for *Penicillium* *[1] [2 max]*

22.6/22.7

1 Lack of nutrients; *[1]*

   Lack of oxygen; *[1]*

   Temperature that is too high or low; *[1]*

   Build-up of toxic waste products; *[1]*

   Incorrect pH *[1] [3 max]*

2 Population is allowed to reach stationary phase; *[1]*

   Secondary metabolites are produced; *[1]*

   Penicillin is a secondary metabolite; *[1]*

   Penicillin is produced as a defence mechanism *[1] [max 3]*

3 2.8×10^5 *[4]*

   If the final answer is incorrect, marks can be scored (up to a maximum of 3) for the following:

   Dilution factor of 20 000; *[1]*

   Estimate from plate 1 = 320 000; *[1]*

   Estimate from plate 2 = 240 000; *[1]*

   Mean = 280 000; *[1]*

22.8

1 Within a polysaccharide/cellulose/gelatin matrix; *[1]*

   One side of a semi-permeable membrane *[1]*

2 Set-up tends to be more expensive than the use of enzymes free in solution; *[1]*

   Less downstream processing, which saves money; *[1]*

   Continuous use increases efficiency and production rate; *[1]*

The balance of costs and benefits may depend on the type of reaction; [1]

For established biotechnology uses, the economic benefits outweigh the costs [1] [3 max]

3 Semi-synthetic penicillin has a slightly different structure to natural penicillin; [1]

Less bacterial resistance to semi-synthetic penicillin [1]

23.1/23.2

1 Inedible / indigestible parts; [1]

Energy used in respiration / to produce heat; [1]

Energy used for movement; [1]

Some energy lost in excretion [1] [max 3]

2 Restrict movement; [1]

Use antibiotics to reduce the amount of energy used by the immune system; [1]

Maintain optimal environmental temperature; [1]

More energy is converted to biomass [1] [max 3]

3 a $0.1\,kg\,m^{-2}\,yr^{-1}$ [2]

b (Genetic modification of the crop could increase) resistance to disease/pests; [1]

Herbicide resistance; [1]

Drought resistance; [1]

This enables increased growth and higher yields [1] [max 3]

23.3

1 Nitrogen fixation; [1]

ammonification; [1]

nitrite; [1]

denitrification [1]

2 Deforestation; [1]

reduces photosynthesis; [1]

More combustion; [1]

in industry/vehicles/as fuel; [1]

Higher carbon dioxide concentrations in the atmosphere [1] [max 3]

3 Anaerobic conditions; [1]

Increased denitrification; [1]

Reduced nitrification; [1]

Lower concentration of nitrates for crop plants to absorb [1] [max 3]

23.4

1 (A plagioclimax is) a community resulting from deflected succession, where human influence has prevented succession from producing a climax community; [1]

(Examples include) managed forests, grazing grassland, arable crop fields [1]

2 Tolerance of extreme conditions/environments; [1]

The ability to fix nitrogen from the atmosphere; [1]

The ability to photosynthesise; [1]

The production of many seeds/spores that can be carried by wind; [1]

Rapid germination of seeds [1][max 3]

3 Primary succession is the natural development of an ecosystem to form a climax community; [1]

Deflected succession is when human activity interrupts succession, producing a plagioclimax; [1]

Primary succession tends to produce greater biodiversity because deflected succession reduces the number of ecological niches available [1]

23.5

1 96 000 [2]

[If the final answer is incorrect, '1 000 000 / 500' scores 1 mark]

2 Different areas within an ecosystem are identified; [1]

These areas vary in their abiotic conditions; [1]

Random sampling may miss some of the different areas [1] [max 2]

3 43 [2] [1 mark for 42.66]

24.1

1 a A factor that limits the maximum size of a population; [1]

b A factor that affects population size, but its effects vary with population density [1]

2 Birth rate is always higher than death rate; [1]

An increase in population size, with a relatively constant rate of change for much of the two stages; [1]

The increase occurs because death rate declines at an earlier stage than birth rate; [1]

This means the difference in the two rates widens during these stages [1] [max 3]

3 Country A has the faster population growth; [1]

A greater proportion of its population can be found in the younger age categories [1]

24.2/24.3

Go further: Allelopathy – competition between plants

1 When leaves are eventually shed and decompose, the allelopathic chemicals will diffuse through the soil.

2 Allelopathic chemicals may be persistent and remain in the soil for a long time, therefore interfering with the growth of crop plants.

Summary questions

1 Light; *[1]*

Water (from soil); *[1]*

Carbon dioxide; *[1]*

(Named) mineral ions (from soil) *[1] [3 max]*

2 The filter feeders have the same ecological niche; *[1]*

They feed on similar food items; *[1]*

Barnacles avoid competition; *[1]*

The competition is interspecific *[1] [3 max]*

3 (The predator peaks are lower because) there are fewer predators in the ecosystem; *[1]*

(The peaks are delayed because) when prey numbers rise, the chance of survival increases for predators; *[1]*

Intraspecific competition is reduced; *[1]*

After a short time, predator numbers rise; *[1]*

Reproduction and the rearing of offspring contribute to the delay *[1] [max 3]*

24.4/24.5

1 Conservation is active/manages an ecosystem/ requires interference; *[1]*

Preservation avoids interference/aims to leave an ecosystem undisturbed *[1]*

2 Tilapia are lower down the food chain; *[1]*

Only a single energy transfer occurs (from producer to primary consumer); *[1]*

Farming salmon requires the depletion of wild (primary/secondary consumer) fish populations to act as food for the salmon *[1] [2 max]*

3 Sometimes it is unclear which species were present in the original community; *[1]*

Understanding and restoring the original abiotic conditions may be difficult; *[1]*

Succession takes a long time *[1][2 max]*

24.6–24.9

1 An ecosystem that is especially vulnerable to environmental change/human interference *[1]*

2 Human populations are no longer nomadic; *[1]*

The same land is grazed/deforested for long periods of time; *[1]*

Vegetation is unable to recover; *[1]*

The risk of soil erosion increases *[1] [3 max]*

3 Preservation is usually impossible because so few peat bogs remain in their original state; *[1]*

Peat bogs take a long time to form; *[1]*

There is a conflict between reclamation/ conservation and human activity (e.g. extraction of peat) *[1]*

▼ *Table of values of* t

Degree of freedom (df)	p values			
	0.10	0.05	0.01	0.001
1	6.31	12.71	63.66	636.60
2	2.92	4.30	9.92	31.60
3	2.35	3.18	5.84	12.92
4	2.13	2.78	4.60	8.61
5	2.02	2.57	4.03	6.87
6	1.94	2.45	3.71	5.96
7	1.89	2.36	3.50	5.41
8	1.86	2.31	3.36	5.04
9	1.83	2.26	3.25	4.78
10	1.81	2.23	3.17	4.59
12	1.78	2.18	3.05	4.32
14	1.76	2.15	2.98	4.14
16	1.75	2.12	2.92	4.02
18	1.73	2.10	2.88	3.92
20	1.72	2.09	2.85	3.85
α	1.64	1.96	2.58	3.29

▼ *Critical values for Spearman's rank correlation coefficient,* r_s

	$p = 0.1$	$p = 0.05$	$p = 0.02$	$p = 0.01$			$p = 0.1$	$p = 0.05$	$p = 0.02$	$p = 0.01$
	5%	$2\frac{1}{2}$%	1%	$\frac{1}{2}$%	1-Tail Test		5%	$2\frac{1}{2}$%	1%	$\frac{1}{2}$%
	10%	5%	2%	1%	2-Tail Test		10%	5%	2%	1%
n						n				
1	–	–	–	–		31	0.3012	0.3560	0.4185	0.4593
2	–	–	–	–		32	0.2962	0.3504	0.4117	0.4523
3	–	–	–	–		33	0.2914	0.3449	0.4054	0.4455
4	1.0000	–	–	–		34	0.2871	0.3396	0.3995	0.4390
5	0.9000	1.0000	1.0000	–		35	0.2829	0.3347	0.3936	0.4328
6	0.8286	0.8857	0.9429	1.0000		36	0.2788	0.3300	0.3882	0.4268
7	0.7143	0.7857	0.8929	0.9286		37	0.2748	0.3253	0.3829	0.4211
8	0.6429	0.7381	0.8333	0.8810		38	0.2710	0.3209	0.3778	0.4155
9	0.6000	0.7000	0.7833	0.8333		39	0.2674	0.3168	0.3729	0.4103
10	0.5636	0.6485	0.7455	0.7939		40	0.2640	0.3128	0.3681	0.4051
11	0.5364	0.6182	0.7091	0.7545		41	0.2606	0.3087	0.3636	0.4002
12	0.5035	0.5874	0.6783	0.7273		42	0.2574	0.3051	0.3594	0.3955
13	0.4835	0.5604	0.6484	0.7033		43	0.2543	0.3014	0.3550	0.3908
14	0.4637	0.5385	0.6264	0.6791		44	0.2513	0.2978	0.3511	0.3865
15	0.4464	0.5214	0.6036	0.6536		45	0.2484	0.2945	0.3470	0.3822
16	0.4294	0.5029	0.5824	0.6353		46	0.2456	0.2913	0.3433	0.3781
17	0.4142	0.4877	0.5662	0.6176		47	0.2429	0.2880	0.3396	0.3741
18	0.4014	0.4716	0.5501	0.5996		48	0.2403	0.2850	0.3361	0.3702
19	0.3912	0.4596	0.5351	0.5842		49	0.2378	0.2820	0.3326	0.3664
20	0.3805	0.4466	0.5218	0.5699		50	0.2353	0.2791	0.3293	0.3628